Learning Guide for

Principles of Human Anatomy

Eighth Edition

Gerard J. Tortora
Bergen Community College

Robert J. Amitrano
Bergen Community College

An imprint of Addison Wesley Longman, Inc.

Don Mills, Ontario ■ Sydney ■ Mexico City ■ Madrid ■ Amsterdam
Menlo Park, California ■ Reading, Massachusetts ■ New York
Bonn ■ Paris ■ Milan ■ Singapore ■ Tokyo ■ Seoul ■ Taipei

Sponsoring Editor: Kay Ueno
Publishing Assistant: Nikki Richesin
Senior Production Editor: Larry Olsen
Composition: Scott Silva
Cover Designer: Yvo Riezebos

ISBN 0-321-03660-3

1 2 3 4 5 6 7 8 9 10—CRS—03 02 01 00 99 98

Benjamin/Cummings Publishing
2725 Sand Hill Road
Menlo Park, California 94025

Contents

Preface

This Learning Guide accompanies Gerard J. Tortora's *Principles of Human Anatomy,* Eighth Edition. It is designed to allow you to gain mastery over the subject material through active learning rather than rote memorization.

Each of the 27 chapters of this Learning Guide correlates with the information in each chapter of *Principles of Human Anatomy,* Eighth Edition. **Topic Outline** and **Objectives** parallel the text organization and present learning goals for each chapter. **Scientific Terminology** lists the prefixes and suffixes associated with the nominclature found in the corresponding text chapter. The **Study Questions** provide you with a myriad of learning activities correlated to topics in each chapter and cross-references to text pages. You can check your progress via **Answers to Select Questions,** which accompany each chapter. At the conclusion of each chapter is a **Self Quiz t**hat provides a final check of subject mastery. **Answers to the Self Quiz** immediately follow the Self Quiz.

A unique change to the Eighth Edition of *Principles of Human Anatomy* is the addition of a new booklet entitled *Applications to Health and Cross Reference Guide to A.D.A.M.*® *Interactive Anatomy.* Several chapters of this Learning Guide contain pertinent questions relating to this booklet. These questions and the location of their answers have been duly noted in each chapter of this guide.

This Learning Guide was developed with the student in mind. No two students are alike; each has different strengths and weaknesses when it comes to the learning process. Some students study alone while others prefer being part of a study group; some like multiple-choice and fill-in questions while others like essay questions. The choices of activities included in each chapter are based on learning exercises that I have used during my past ten years of teaching allied health students at Bergen Community College. Based upon this experience, I believe that this Learning Guide will satisfy the needs of everyone using it.

I would like to thank my wife, Suzeanne, and my son, Robert, for their patience and under-standing while I was revising this Learning Guide. I would also like to thank my good friend and colleague, Professor Gerard J. Tortora, for his support and guidance during the revision of this guide. In addition, I would like to thank my current and former students for allowing me to test the material on them.

Robert J. Amitrano
Bergen Community College

An Introduction to the Human Body

CHAPTER 1

SYNOPSIS

In the 1966 film *Fantastic Voyage,* a group of surgeons and scientists were miniaturized and injected into a body in order to perform a delicate brain operation. This futuristic film attempted to give the viewer a glimpse into the wonders of the human body.

You are about to commence your own "fantastic voyage." As you embark on what will hopefully be a long and enjoyable journey you will discover the magnificence of this wondrous machine called the human body.

In the same way that a mechanic must completely understand the component parts of an automobile, so too must the student of anatomy have a solid comprehension of the structures that comprise the human body. In this chapter you will be introduced to the **levels of structural organization** and the **life processes** and receive an overview of the human body. Topics in this discussion include **anatomical terminology** and position, the **structural plan**, the usage of **directional terms**, **planes** and **sections of the body**, **body cavities**, and the **abdominal regions** and **quadrants**.

It is from this basic knowledge of chemical, cellular, tissue, organ, and system organization that you will begin your journey. In the following chapters, you will build upon this foundational information.

TOPIC OUTLINE AND OBJECTIVES

A. Anatomy Defined

1. Define anatomy, with its subdivisions and physiology.

B. Levels of Structural Organization

2. Define and explain the levels of structural organization that comprise the human body: chemical, cellular, tissue, organ, system, and organismic.
3. Identify, define, and describe the function of each body system.

C. Life Processes

4. List and define the six life processes.

D. Overview of the Human Body

5. Describe the structural plan of the human body.
6. Define anatomical position, and compare the common and anatomical names of the various body regions.
7. Define and discuss the usage of directional terms.
8. Define the planes and sections of the human body, and explain how they are made.
9. Name and locate the principal body cavities and the organs within them.
10. Contrast how the abdominopelvic regions and quadrants are produced.

E. Medical Imaging

F. Measuring the Human Body

SCIENTIFIC TERMINOLOGY

Find an anatomical sample word for each prefix and suffix:

Prefix/Suffix	Meaning	Sample Word
ana-	upward	
auto-	self	
cardi-	heart	
cata-	downward	
chondro-	cartilage	
cyto-	cell	
epi-	above	
-graph	to write	
histo-	tissue	
hypo-	under	
-logos	to study	
media-	middle	
meta-	change	
para-	near	
patho-	disease	
peri-	around	
pleur-	rib, side	
quad-	four	
radio-	ray	
-tome	cut	

A. Anatomy Defined (page 2)

A1. _____ refers to the study of structure and the relationships among structures.

A2. Cytology describes the chemical and microscopic study of the structure of a

_____.

A3. The study of development from the fertilized egg to adult form is called

_____.

A4. _____ refers to the study of structures that can be examined without the use of a microscope.

B. Levels of Structural Organization (pages 2–5)

B1. Complete the following questions about the levels of organization.

a. The basic structural and functional living units of an organism are found on the

_____ level.

b. Tissues are groups of _____ _____ together with their

_____ _____ (substance between cells).

c. Which level of organization is best described by the joining together of two or more

different tissues? _____

d. The four basic types of tissues in the body are:

1. _____

2. _____

3. _____

4. _____

e. The highest level of organization is the _____ level.

B2. Check your knowledge of the major body systems by answering these questions.

a. The *(integumentary? respiratory?)* system supplies oxygen, eliminates carbon dioxide, and helps regulate acid–base balance.

b. The spleen, thymus, and tonsils are organs associated with the *(endocrine? lymphatic and immune?)* system.

c. The *(urinary? cardiovascular?)* system distributes oxygen and nutrients and carries carbon dioxide and wastes.

d. The *(digestive? muscular?)* system participates in bringing about movement, maintenance of posture, and heat production.

e. The organs associated with the reproductive system include *(testes and ovaries? kidneys and bladder?)*.

f. Skin, hair, and nails are associated with the *(integumentary? nervous?)* system.

g. The *(muscular? skeletal?)* system supports and protects the body, stores minerals and energy, assists in movement, and houses cells that produce blood cells.

h. Control and integration of body activities via chemical messengers in the blood is the function of the *(lymphatic? endocrine?)* system.

i. The *(nervous? muscular?)* system regulates body activities through action potentials, by detecting changes in the internal and external environment, interpreting changes, and responding to the changes.

C. Life Processes   (pages 6–8)

C1. The breakdown of food into molecules which provide the energy needed to sustain life is referred to as *(anabolism? catabolism?)*.

C2. The process whereby unspecialized cells become specialized cells is *(responsiveness? differentiation?)*.

C3. *(Growth? Reproduction?)* occurs for growth, repair, or replacement, or the production of new cells.

C4. The contraction of the gallbladder to release bile, which aids in the digestion of fats, refers to this process. *(catabolism? movement?)*

C5. An example of *(responsiveness? metabolism?)* would be a nerve cell generating electrical signals known as nerve impulses.

D. Overview of the Human Body (pages 8–16)

D1. Which of the following statements about anatomical position is NOT true?

a. Subject stands erect facing the observer.

b. Arms are placed at the sides.

c. Palms are turned backward.

d. Feet are flat on the floor.

D2. With reference to the structural plan of the body, what is meant by a tube-within-a-tube construction?

D3. Using Figure 1.2 in the textbook, complete the following table of common and anatomical names.

Common Name	Anatomical Term
a.	Cervical
b. Forearm	
c.	Axillary
d. Cheek	
e.	Orbital
f. Thigh	
g.	Pedal
h. Groin	
i.	Acromial
j. Chin	
k.	Crural
l. Breast	
m.	Carpal
n. Arm	
o.	Calcaneal
p. Palm	
q.	Tarsal
r. Fingers	
s.	Gluteal
t. Back of knee	
u. Hand	
v. Sole	
w.	Olecranal
x. Hip	
y.	Oral

D4. Using the directional terms in Table 1.3 and Figure 1.3 in your textbook, fill in the blanks with the correct term.

a. The carpals are _____ to the elbow.

b. The ears are _____ to the nose.

c. The spleen and liver are _____.

d. The heart is located _____ to the diaphragm.

e. Upon arriving at the scene of an accident, you find a body lying face down. The body is

lying on which surface? _____

f. The knee is _____ to the tarsals.

g. The gallbladder and liver are _____.

h. The stomach is located _____ to the heart.

i. The muscles of your arm are _____ to the skin.

j. The vertebral column is located on the _____ side of the body.

k. The ulna is on the _____ side of the forearm.

l. The _____ pleura forms the inner layer of the pleural sacs and covers the external surface of the lungs.

D5. Refer to Figure LG 1.1 and identify each plane listed below.

_____ transverse plane   _____ parasagittal plane

_____ midsaggital plane   _____ frontal plane

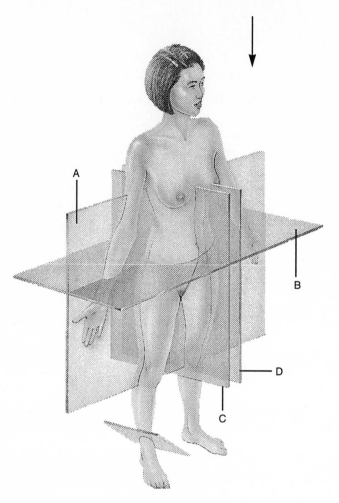

Right anterolateral view

Figure LG 1.1   Planes of the human body.

D6. After reviewing Figures 1.6 and 1.7 in your textbook, complete the following questions pertaining to body cavities.

a. The _____ cavity contains the stomach, spleen, and liver.

b. The urinary bladder, portions of the large intestines, and internal reproductive organs are

located in the _____ cavity.

c. The cavity, formed by the cranial bones, that contains the brain is the

_____ cavity.

d. The pericardial and pleural cavities and the mediastinum are subdivisions of the

_____ cavity.

e. Together, the thoracic and abdominopelvic cavities form the _____ cavity.

D7. Complete Figure LG 1.2 according to the directions below.

a. Select different colors for the cranial, vertebral, thoracic, and abdominopelvic cavities and shade in each cavity. (Make sure that the color code oval corresponds with the color you shade the cavity.)

○ cranial　　　　○ vertebral　　　　○ thoracic　　　　○ abdominopelvic

b. Label the cavities below.

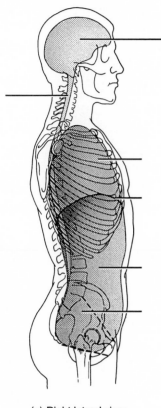

(a) Right lateral view

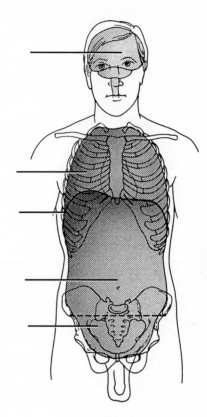

(b) Anterior view

Figure LG 1.2　Body cavities.

D8.　The spinal cord and the beginning of the spinal nerves are located in this cavity:

_____.

D9.　Which organs are located in the following cavities?

a. Pericardial

b. Pleural

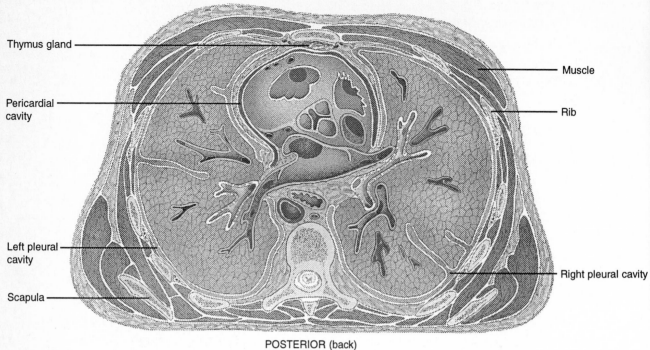

ANTERIOR (front)

Thymus gland

Muscle

Pericardial cavity

Rib

Left pleural cavity

Scapula

Right pleural cavity

POSTERIOR (back)

Transverse section

Figure LG1.3   Mediastinum.

D10.   Refer to Figure LG1.3 and follow the directions below.

a.   Using colored pencils, shade in the following structures. (Make sure that the color code oval corresponds with the color you shade the cavity.)

○   esophagus
○   heart
○   left lung
○   right lung
○   sternum
○   thoracic aorta
○   vertebra

D11.   You are the attending physician when a patient is brought to the emergency room with a knife wound in the epigastric region. Since you have a thorough knowledge of Figure 1.9, you would suspect injury of certain organs. Which organs have possibly been injured?

D12.   Early one morning you develop exquisite tenderness in the right lower quadrant. In addition, you discover that you are running a fever. Give a probable diagnosis for your symptoms.

D13. Name the four lines that border the umbilical region and describe where they are drawn (refer to Fig. 1.9 in the textbook).

E. Medical Imaging (page 16)

E1. Refer to Table 1.5 and match the medical imaging with the correct definition.

CR	conventional radiography
CT	computerized tomography
DSA	digital subtraction angiography
DSR	dynamic spatial reconstruction
MRI	magnetic resonance imaging
PET	positron emission tomography
US	ultrasound

a. _____ Provides information on function as well as structure.

b. _____ A single barrage of x-rays passes through the body and produces a two-dimensional image of the interior of the body.

c. _____ High-frequency sound waves produced by a hand-held transducer reflect back to produce an image on a video monitor.

d. _____ Image can be rotated, tipped, "sliced open," enlarged, replayed, and viewed in slow and high-speed motion.

e. _____ X-ray source arcs around the body, producing a cross-sectional picture.

f. _____ A computer compares an x-ray image of a region of the body before and after a contrast dye has been introduced.

g. _____ Noninvasive and uses no radiation, but is not indicated for pregnant women or persons with pacemakers or metal joints.

ANSWERS TO SELECT QUESTIONS

A1. Anatomy.

A2. Cell.

A3. Developmental anatomy.

A4. Gross (macroscopic) anatomy.

B1. (a) Cellular; (b) similar cells, intercellular substance; (c) organs; (d) 1. epithelial tissue, 2. connective tissue, 3. muscle tissue, 4. nervous tissue; (e) organismic.

B2. (a) Respiratory; (b) lymphatic; (c) cardiovascular; (d) muscular; (e) testes and ovaries; (f) integumentary; (g) skeletal; (h) endocrine; (i) nervous.

C1. Catabolism.

C2. Differentiation.

C3. Reproduction.

C4. Movement.

C5. Responsiveness.

D1. (c).

D2. Outer tube is formed by the body wall; inner tube is part of the gastrointestinal tract.

D3.

Common Name	Anatomical Term
a. **Neck**	Cervical
b. Forearm	**Antebrachial**
c. **Armpit**	Axillary
d. Cheek	**Buccal**
e. **Eye**	Orbital
f. Thigh	**Femoral**
g. **Foot**	Pedal
h. Groin	**Inguinal**
i. **Shoulder**	Acromial
j. Chin	**Mental**
k. **Leg**	Crural
l. Breast	**Mammary**
m. **Wrist**	Carpal
n. Arm	**Brachial**
o. **Heel**	Calcaneal
p. Palm	**Metacarpal**
q. **Ankle**	Tarsal
r. Fingers	**Phalangeal**
s. **Buttocks**	Gluteal
t. Back of knee	**Popliteal**
u. Hand	**Manual**
v. Sole	**Plantar**
w. **Back of elbow**	Olecranal
x. Hip	**Coxal**
y. **Mouth**	Oral

D4. (a) Distal; (b) lateral; (c) contralateral; (d) superior; (e) anterior; (f) proximal; (g) ipsilateral; (h) inferior; (i) deep; (j) posterior; (k) medial; (l) visceral.

D5. (a) Frontal; (b) transverse; (c) midsagittal; (d) parasagittal.

D6. (a) Abdominal; (b) pelvic; (c) cranial; (d) thoracic; (e) anterior.

D8. Vertebral (spinal) canal.

D9. (a) Heart; (b) lungs.

D11. Stomach, liver, transverse colon, pancreas, duodenum, abdominal aorta, inferior vena cava, left kidney.

D12. Appendicitis.

D13. Right and left midclavicular lines—drawn slightly medial to the nipples. Subcostal line—drawn across the lowest border of the rib cage. Transtubercular line—drawn across the iliac crests.

E1. (a) PET; (b) CR; (c) US; (d) DSR; (e) CT; (f) DSA; (g) MRI.

SELF QUIZ

Choose the one best answer to the following questions.

1. The study of structural changes associated with disease is

   A. histology
   B. cytology
   C. embryology
   D. radiography
   E. pathology

2. The next higher level of organization above the tissue level is

   A. cellular
   B. chemical
   C. organ
   D. system
   E. organismic

3. The process of an unspecialized cell changing into a specialized cell is called

   A. growth
   B. reproduction
   C. differentiation
   D. responsiveness
   E. metabolism

4. Which system is best described by these functions: elimination of wastes, secretion of a hormone that helps regulate fluid and electrolyte balance and volume, and regulation of red blood production?

   A. respiratory
   B. cardiovascular
   C. endocrine
   D. lymphatic
   E. urinary

5. Which of the following terms best describes the relationship of the skin of the chest to the ribs?

   A. anterior
   B. posterior
   C. superficial
   D. deep
   E. medial

6. The palm of the hand is referred to as

   A. plantar
   B. popliteal
   C. sural
   D. crural
   E. metacarpal

7. The plane that divides the body into anterior and posterior halves is called the

   A. sagittal
   B. transverse
   C. horizontal
   D. coronal
   E. cross-sectional

8. What position is the body in when the anterior side is lying down?

   A. posterior
   B. medial
   C. dorsal
   D. prone
   E. supine

9. The study of structures without the use of a microscope is called _____ anatomy.

   A. surface
   B. systemic
   C. gross
   D. developmental
   E. radiographic

10. Which system helps to regulate body temperature, eliminates wastes, helps to synthesize vitamin D, and receives certain stimuli?

    A. muscular
    B. urinary
    C. endocrine
    D. lymphatic
    E. integumentary

11. Any body part that is closer to the midline is said to be

    A. anterior
    B. medial
    C. intermediate
    D. lateral
    E. sagittal

12. The gallbladder is located in which region?

    A. left lumbar
    B. right lumbar
    C. epigastric
    D. left hypochondriac
    E. right hypochondriac

13. The term parasagittal indicates which of the following?

    A. equal left and right halves
    B. equal superior and inferior halves
    C. unequal left and right halves
    D. unequal superior and inferior halves
    E. unequal anterior and posterior halves

14. Which body cavity contains the urinary bladder and internal reproductive organs?

    A. vertebral
    B. abdominal
    C. pelvic
    D. thoracic
    E. pleural

15. Which structures are located in the left iliac region?

    A. stomach and spleen
    B. pancreas and spleen
    C. liver and gallbladder
    D. ascending colon and liver
    E. junction of descending and sigmoid parts of colon

Answer (T) True or (F) False to the following questions.

16. _____ The endocrine system includes the spleen, thymus, and tonsils.

17. _____ The esophagus is anterior to the trachea.

18. _____ The back of the elbow is also known as the olecranon.

19. _____ The transtubercular line is the superior of the two horizontal lines used to create the abdominopelvic region.

20. _____ Structures located on the same side of the body are said to be ipsilateral.

21. _____ The lungs are located within the pericardial cavity.

Fill in the blanks.

22. The heart, thymus gland, esophagus, and trachea are located in the _____.

23. The carpals are _____ to the phalanges.

24. The _____ system breaks down and absorbs food for use by the cells.

25. _____ refers to an increase in size or number.

ANSWERS TO THE SELF QUIZ

1. E	10. E	19. F
2. C	11. B	20. T
3. C	12. E	21. F
4. E	13. C	22. Mediastinum
5. C	14. C	23. Proximal
6. E	15. E	24. Digestive
7. D	16. F	25. Growth
8. D	17. F	
9. C	18. T	

Cells

CHAPTER

2

SYNOPSIS

Millions of people annually travel to the Empire State Building in New York City to partake of the breathtaking view that can be seen from the observation platform. Few, if any, of these visitors ever contemplate the tons of materials that were required to build this magnificent structure. The human body, likewise, is composed of nearly 100 trillion cells classified into 200 different cell types, and yet few individuals ever consider the contribution of these splendid structures.

The **cell** is the basic structural and functional unit of the body, and its integrity is essential to life. One should not, however, assume that the cell is a mere building block of the body; rather, it is an extraordinary structure in and of itself. Bound by a floating, ever-moving membrane, the cell contains an internal architecture that can not be rivaled.

Semifluid **cytosol**, surrounded by the plasma membrane, provides the environment for many of the cell's chemical reactions while at the same time supporting the **organelles** and **inclusions** of the cell.

The **organelles** (little organs) function in much the same manner that the organs of our bodies do. Each has its own specific structure and function: some produce proteins, others are involved in packaging and secretion, and others produce energy. Nevertheless, it is their precise, coordinated actions that enable a cell to perform the specific task for which it was designed.

As your foundational learning continues you will be presented with the structural and functional aspects of a generalized cell. Cellular components will be detailed along with the important processes of **passive** and **active transport**, and **normal cellular reproduction**. This chapter also includes a discussion of **abnormal cell division**, **aging**, and key medical terms.

TOPIC OUTLINE AND OBJECTIVES

A. Generalized Cell

1. Examine a cell and its four principal parts.

B. Plasma (Cell) Membrane

2. Describe the chemistry and anatomy of the plasma membrane.
3. Discuss the four functions associated with the membrane.

4. Contrast how materials cross the plasma membrane by diffusion, facilitated diffusion, osmosis, filtration, active transport, endocytosis, and exocytosis.

C. Cytoplasm

5. Describe the structure and function of the cytoplasm.

D. Organelles

6. Describe the structure and function of the following organelles: nucleus, ribosomes, endoplasmic reticulum (ER), Golgi complex, lysosomes, peroxisomes, mitochondria, the cytoskeleton, centrosome, cilia, and flagella.

E. Cell Inclusions

7. Define a cell inclusion and give several examples.

F. Normal Cell Division

8. Contrast stages and outcome of mitosis and meiosis.
9. **Describe apoptosis.**

G. Cells and Aging

10. Explain the relationship between aging and cells.

H. Cancer

11. Describe cancer (CA) as a type of abnormal cell division.

I. Key Medical Terms Associated with Cells

SCIENTIFIC TERMINOLOGY

Find an anatomical sample word for each prefix and suffix:

Prefix/Suffix	Meaning	Sample Word
auto-	self	
chromo-	color	
cilio-	eyelashes	
diffus-	spreading	
endo-	into, within	
extra-	outside	
exo-	out of	
hyper-	over	
inter-	between	
intra-	inside, within	
-logos	to study	
-lysis	dissolution	
meta-	after	
mitos-	thread	
neo-	new	
onco-	swelling, mass	
-osis	process	
perox-	peroxide	
phago-	eating	
plasmic	cytoplasm	
-pous	foot	
pseudo-	false	
soma-	body	
-stasis	standing still	
telos-	far or end	

A. Generalized Cell   (pages 29–30)

A1.   List the four principal parts of a generalized cell.

a.   _____

b.   _____

c.   _____

d.   _____

B. Plasma (Cell) Membrane   (pages 30–33)

B1.   Answer the following questions about cellular chemistry and function.

a.   Why is the plasma (cell) membrane described as a fluid mosaic?

b.   What is meant by the term phospholipid bilayer when describing the plasma (cell) membrane?

c.   Differentiate between the functions of integral and peripheral proteins.

Integral   **Peripheral**

B2.   List and discuss the four functions of the plasma (cell) membrane.

a.

b.

c.

d.

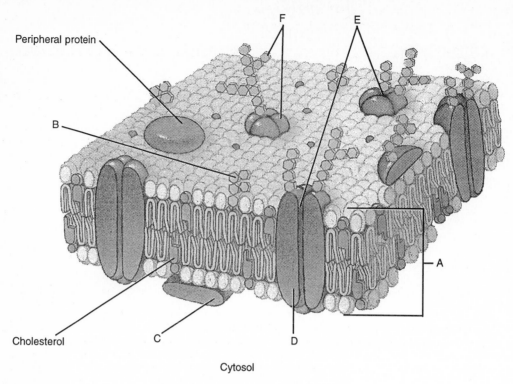

Extracellular fluid

Peripheral protein

B

Cholesterol

C

D

Cytosol

Details of plasma membrane structure

Figure LG 2.1 Plasma membrane.

B3. Label parts A–F of the plasma membrane in Figure LG 2.1.

B4. Test your understanding of the movement of materials across a plasma membrane.

a. _____ is a passive process in which there is a net movement of molecules or ions from a region of their higher concentration to a region of their lower concentration.

b. The process involving the movement of solvents, such as water, and dissolved substances across a selectively permeable membrane by gravity or mechanical pressure describes

_____.

c. This passive process is accomplished with the assistance of integral proteins in the

membrane that serve as carriers: _____ _____.

d. _____ is the net movement of water molecules through a selectively permeable membrane from an area of high water concentration to an area of low water concentration.

e. Active transport is a process by which substances are transported across a plasma

membrane from an area of their _____ concentration to an area of their

_____ concentration.

f. The source of energy used for active transport is _____.

g. Name the three basic kinds of endocytosis:

_____, _____, _____.

h. A _____ _____ is a membrane sac that develops as pseudopodia surround and engulf a solid particle.

i. _____ _____ is similar to pinocytosis; however, it is a highly specific process in which cells can take up specific molecules or particles.

C. Cytoplasm (pages 33–34)

C1. List the chemical constituents of cytoplasm.

D. Organelles (pages 34–44)

D1. Complete the table below by naming the organelle or giving a description of its function.

Organelles	Functions
a.	Contains genes and controls cellular activity.
Ribosomes	b.
c.	Sites of production of most ATP.
Peroxisome	d.
e.	Digest molecules and foreign microbes.
Centrosome	f.
g.	Allow movement of entire cell or movement of a substance along surface of cell.
Golgi complex	h.
Intermediate filaments	i.
j.	Form part of the cytoskeleton; involved in muscle fiber contraction; provide support and shape.
Microtubules	k.
Endoplasmic reticulum	l.

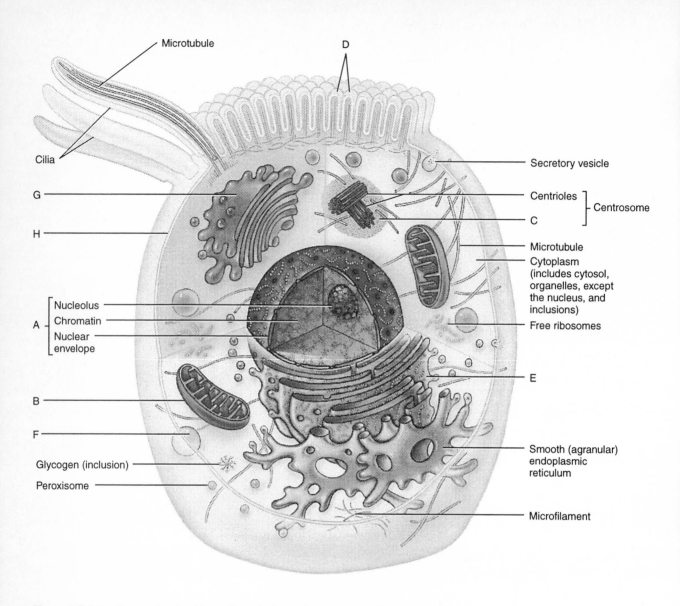

Figure LG 2.2 Generalized cell based on electron microscope studies.

D2. Label parts A–H of the generalized cell in Figure LG 2.2.

E. Cell Inclusions (page 44)

E1. Match the following cellular inclusions with the correct description.

G glycogen	M melanin	T triglycerides

a. _____ Are stored in adipocytes, may decompose to synthesize ATP.

b. _____ A polysaccharide that is stored in the liver, skeletal muscle, and uterine and vaginal mucosa.

c. _____ Pigment stored in certain cells of skin, hair, and eyes.

F. Normal Cell Division   (pages 44–49)

F1.   Match the following terms with their description (two answers apply to each).

MI mitosis	RD reproductive division
ME meiosis	SC somatic cell division

a. _____ Produce two daughter cells identical to the parent cell.

b. _____ Mechanism by which egg and sperm cells are produced.

F2.   Match the stage of mitosis with the corresponding activity.

A anaphase	C cytokinesis
M metaphase	P prophase
T telophase	

a. _____ Cleavage furrow forms around center of cell, progresses inward, and separates cytoplasm into two separate and usually equal portions.

b. _____ Centromeres divide and identical sets of chromosomes move to opposite poles of cell.

c. _____ Chromatin shortens and coils into chromosomes, nucleoli and nuclear envelope disappear, a centrosome and its centrioles move to opposite poles, and mitotic spindles form.

d. _____ Centromeres of chromatid pairs line up on metaphase plate of cell.

e. _____ Nuclear envelope reappears and encloses chromosomes, chromosomes resume chromatin form, nucleoli reappear, and mitotic spindle disappears.

F3.   Complete this question pertaining to apoptosis. Apoptosis (genetically programmed cell death) occurs postnatally to:

a. _____

b. _____

G. Cells and Aging   (pages 50–51)

G1.   List the cells that are incapable of replacement.

G2.   Free radicals are electrically charged molecules that have an unpaired electron. Such

molecules are unstable and can easily damage nearby _____,

_____, and _____.

H. Cancer   (Refer to *Applications to Health* Companion to answer Part H questions.)

H1. Match the following terms with their definition.

> C   cancerous   NC   noncancerous

a. _____ benign growth   b. _____ malignant growth

H2. Match the tumor type to its definition.

> C   carcinoma   M   melanoma
> L   leukemia   O   osteogenic sarcoma
> LY lymphoma   S   sarcoma

a. _____ Cancer of blood-forming organs characterized by rapid white blood cell growth with distorted development.

b. _____ Cancer arising in connective tissues or muscle cells.

c. _____ Malignant disease of lymphatic tissue.

d. _____ Cancer of the bone.

e. _____ Cancerous growth of melanocytes.

f. _____ A malignant tumor arising from epithelial cells.

H3. List two major suspected triggers for cancer development.

a. _____

b. _____

ANSWERS TO SELECT QUESTIONS

A1. (a) Plasma membrane; (b) cytosol; (c) organelles; (d) inclusions.

B1. (a) Proteins floating like icebergs in sea of lipids; (b) membrane has two parallel rows of phospholipid molecules, phosphates facing to the water and fatty acids facing away from the water; (c) integral protein: transporters, channels, receptors, enzymes; peripheral proteins: enzymes, cytoskeletal anchors, cell identity markers.

B2. (a) Communication; (b) shape and protection; (c) electrochemical gradient; (d) selective permeability.

B3. A, phospholipid bilayer; B, glycolipid; C, peripheral protein; D, integral protein; E, channels (pores); F, glycoprotein.

B4. (a) Diffusion; (b) filtration; (c) facilitated diffusion; (d) osmosis; (e) low, high; (f) ATP; (g) phagocytosis, pinocytosis; receptor-mediated endocytosis; (h) phagocytic vesicle; (i) receptor-mediated endocytosis.

C1. Water, proteins, lipids, carbohydrates, inorganic substances.

D1.

Organelles	Functions
a. Nucleus	Contains genes and controls cellular activity.
Ribosomes	**b. Site of protein synthesis.**
c. Mitochondria	Sites of production of most ATP.
Peroxisome	**d. Have enzymes to dispose of hydrogen peroxide.**
e. Lysosome	Digest molecules and foreign microbes.
Centrosome	**f. Organizing center for microtubule formation**
g. Flagella and cilia	Allow movement of entire cell or movement of a substance along surface of cell.
Golgi complex	**h. Package synthesized proteins for secretion; form lysosome.**
Intermediate filaments	**i. Provide structural reinforcement, hold organelles in place**
j. Microfilaments	Form part of the cytoskeleton; involved in muscle fiber contraction; provide support and shape; assist in cell movement and movement within cells.
Microtubules	**k. Form part of the cytoskeleton; form intracellular conducting channels; structural component of cilia, flagella, centrioles and mitotic spindles.**
Endoplasmic reticulum	**l. Surface for chemical reactions; pathway for chemical transportation; synthesis of proteins, detoxifies certain molecules, releases calcium to trigger muscle contraction.**

D2. A, nucleus; B, mitochondrion; C, pericentriolar area; D, microvilli; E, rough endoplasmic reticulum; F, lysosome; G, Golgi complex; H, plasma membrane.

E1. (a) T; (b) G; (c) M.

F1. (a) MI, SC; (b) ME, RD.

F2. (a) C; (b) A; (c) P; (d) M; (e) T.

F3. (a) regulate the number of cells in a tissue; (b) eliminate potentially dangerous cells.

G1. Heart cells, skeletal muscle fibers, nerve cells.

G2. Lipids, proteins, nucleic acids.

H1. (a) NC; (b) C.

H2. (a) L; (b) S; (c) LY; (d) O; (e) M; (f) C.

H3. (a) carcinogens; (b) viruses.

SELF QUIZ

Choose the one best answer to the following questions.

1. The passive movement of glucose through a semipermeable membrane, with the help of an integral protein, is an example of

   A. diffusion
   B. osmosis
   C. bulk flow
   D. facilitated diffusion

2. Which is NOT a function of an integral protein?

   A. serve as a transporter
   B. can form minute channels (pores)
   C. serve as receptor
   D. identify the cell

3. The movement, through a membrane, of water from an area of high concentration to an area of low concentration is referred to as

   A. phagocytosis
   B. pinocytosis
   C. diffusion
   D. osmosis

4. Reversion to a more primitive or undifferentiated form is referred to as

   A. metastasis
   B. atrophy
   C. necrosis
   D. anaplasia

5. Cytoplasm is approximately _____% water.

   A. 40–50
   B. 60–70
   C. 75–90
   D. 90–100

6. Which of the following is NOT found in the nucleus?

   A. nucleolus
   B. chromatin
   C. centrioles
   D. chromosomes

7. If a cell undergoing mitosis normally has 20 chromosomes, how many chromosomes will each daughter cell have?

   A. 10
   B. 20
   C. 40
   D. 80

8. Which of the following is NOT a cellular inclusion?

   A. melanin
   B. lysosome
   C. glycogen
   D. triglycerides

9. This organelle helps to organize the mitotic spindles during cell division.

   A. microfilaments
   B. centrosomes
   C. intermediate filaments
   D. Golgi complex

10. Chromosome replication occurs during which cellular phase?

    A. metaphase
    B. interphase
    C. prophase
    D. anaphase

11. The transformation of one cell type into another is called

    A. hyperplasia
    B. metastasis
    C. dysplasia
    D. metaplasia

12. If you were to spill a bottle of cologne in a classroom, the fact that the fragrance eventually reaches the hallway involves the process of

    _____.

    A. active transport
    B. osmosis
    C. facilitated diffusion
    D. diffusion

13. An increase in the number of cells due to an increase in the frequency of cell division is referred to as

    A. hyperplasia
    B. hypertrophy
    C. metaplasia
    D. anaplasia

14. Which term refers to an increase in the size of a cell?

    A. atrophy
    B. hyperplasia
    C. dysplasia
    D. hypertrophy

Answer (T) True or (F) False to the following questions.

15. _____ Tay-Sachs disease is a rare, inherited disease that causes rapid aging.

16. _____ Ribosomes are involved in protein synthesis.

17. _____ Pinocytosis is considered a passive process.

18. _____ The movement of mucus in the respiratory passageways is a function of the flagella.

19. _____ Nuclear pores in the envelope allow most ions and water-soluble molecules to shuttle between the nucleus and the cytoplasm.

20. _____ Interstitial fluid and intercellular fluid are two different fluids.

ANSWERS TO THE SELF QUIZ

1. D	8. B	15. F
2. D	9. B	16. T
3. D	10. B	17. F
4. D	11. D	18. F
5. C	12. D	19. T
6. C	13. A	20. F
7. B	14. D	

Tissues

SYNOPSIS

In the previous chapter you had the opportunity to discover the wonders of the cell. Cells are highly arrayed units, but they do not function in isolation. Working in concert, similar cells and their intercellular (between the cells) material form **tissues**. **Intercellular material** includes the body fluids and cellular secretions. All organs are composed of two or more of the four principal types of tissues: **epithelial, connective, muscle,** and **nervous.**

As you learned in Chapter 1, there is a direct correlation between anatomy and physiology, and this correlation is best seen in tissues. Each type of tissue possesses a distinct structure and equally distinct function. An example is connective tissue, which consists of widely scattered cells supported by a **matrix of ground substance** and assorted **fibers**. It is this matrix that allows connective tissue to function as a binding, protective, and supportive tissue.

This chapter is primarily concerned with a discussion of the structure, function, and location of the principal tissues. Special emphasis will be given to the relationship of structure and function.

TOPIC OUTLINE AND OBJECTIVES

A. Types of Tissues and Their Origins

1. Define tissues, compare their origins, and classify the tissues of the body into four major types.

B. Cell Junctions

C. Epithelial Tissue

2. Describe the general features of epithelial tissue.
3. Discuss the location and function of epithelium.
4. Explain the classification of lining and covering epithelium.
5. Define a gland and contrast exocrine and endocrine glands.

D. Connective Tissue

6. Describe the general features of connective tissue, including the cells, ground substance, and fibers.
7. Discuss the structure, function, and location of the different types of connective tissue.

E. Membranes

8. Define an epithelial membrane and distinguish among mucous, serous, and synovial membrane.

F. Muscle Tissue

9. Describe the structure and location of the three types of muscle tissue.

G. Nervous Tissue

10. Describe the structure and functions of nervous tissue.

SCIENTIFIC TERMINOLOGY

Find an anatomical sample word for each prefix and suffix:

Prefix/Suffix	Meaning	Sample Word
apo-	from	
cardio-	heart	
chondro-	cartilage	
cut-	skin	
dendro-	tree	
desmos-	bond	
endo-	within	
fibro-	fibers	
-glia	glue	
holos-	entire	
hemi-	half	
histio-	tissue	
kerato-	horny	
lipo-	fat	
macro-	large	
mero-	part	
meso-	middle	
micro-	small	
peri-	around	
squama-	scale	
sub-	under	
-thelium	covering	

A. Types of Tissues and Their Origins (page 58)

A1. Epithelial tissue covers the _____ surfaces; lines

_____ organs, body cavities, and ducts; and forms

_____.

A2. Connective tissue _____ and _____ the body and its

organs; _____ organs together; stores _____

reserves; and provides _____.

A3. Muscular tissue is responsible for _____ and generation of force.

A4. Nervous tissue initiates and _____ action _____ that

_____ body activities.

B. Cell Junctions   (pages 58–61)

B1.   Complete the table with the functions of the following junctions.

Cell Junction	Functions
Tight	a.
Plaque-bearing	b.
Gap	c.

C. Epithelial Tissue   (pages 61–70)

C1.   Name the two types of epithelial tissue.

1. _____

2. _____

C2.   Answer the following questions (T) true or (F) false.

a. _____ Epithelium consists of loosely packed cells with little extracellular substance.

b. _____ Epithelia are avascular.

c. _____ Epithelial cells are arranged in a continuous sheet that may be single- or multi-layered.

d. _____ Epithelium has a low capacity for renewal.

e. _____ Epithelial cells rest on a structure called the basement membrane.

f. _____ Functions of epithelia include protection, filtration, and absorption.

C3. Match the following covering and lining epithelium with the correct description.

P pseudostratified	Si simple	St stratified

a. _____ Has only one layer of cells; some cells do not reach the surface.

b. _____ Arranged in single layer; function in absorption and filtration.

c. _____ Cells stacked in several layers; in areas of wear and tear.

C4.   Complete the table below, pertaining to cell shapes.

Squamous	a.
b.	Cube shaped
Columnar	c.
d.	Varies in shape with distension

A Name _____

   Location _____

   Function _____

B Name _____

   Location _____

   Function _____

C Name _____

   Location _____

   Function _____

D Name _____

   Location _____

   Function _____

E Name _____

   Location _____

   Function _____

F Name _____

   Location _____

   Function _____

Figure LG3.1a–f Diagrams of selected tissue types.

G Name _____

   Location _____

   Function _____

H Name _____

   Location _____

   Function _____

I Name _____

   Location _____

   Function _____

J Name _____

   Location _____

   Function _____

K Name _____

   Location _____

   Function _____

L Name _____

   Location _____

   Function _____

Figure LG3.1g–l Diagrams of selected tissue types, continued.

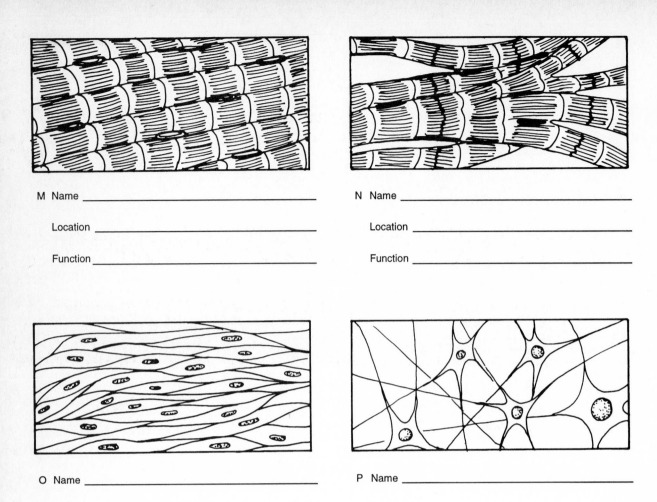

M Name _____

   Location _____

   Function _____

N Name _____

   Location _____

   Function _____

O Name _____

   Location _____

   Function _____

P Name _____

   Location _____

   Function _____

Figure LG3.1m–p   Diagrams of selected tissue types, continued.

C5. After reviewing the diagrams of epithelial tissue types (A–F) in Figure LG 3.1, complete the following questions.

a. Name each tissue type.

b. Give a location(s) for each tissue.

c. List one or more functions.

d. With different colored pencils, shade in the cytoplasm, nucleus, and connective tissue.

C6. Describe the following modifications of epithelial tissue.

a. Cilia

b. Goblet cells

c. Microvilli

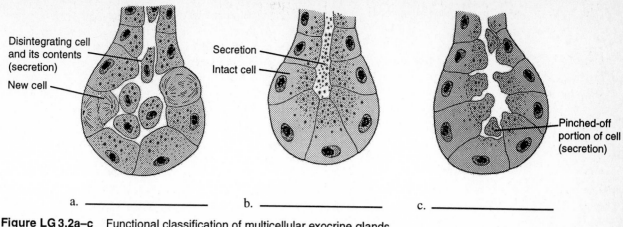

Disintegrating cell
and its contents
(secretion)

New cell

Secretion

Intact cell

Pinched-off
portion of cell
(secretion)

a. _____   b. _____   c. _____

Figure LG 3.2a–c   Functional classification of multicellular exocrine glands.

C7.   Label the parts of Figure LG 3.2 as holocrine, merocrine, or apocrine.

C8.   Differentiate between

a. exocrine glands _____

b. endocrine glands _____

D.   Connective Tissue   (pages 70–81)

D1.   Answer (T) true or (F) false to the following questions about the general features of
connective tissue.

a. _____ Ground substance plus fibers are referred to as the matrix.

b. _____ Connective tissue does occur on free surfaces.

c. _____ All connective tissue is highly vascular.

d. _____ The matrix largely determines the qualities of the connective tissue.

e. _____ All connective tissue has a nerve supply.

D2. List the four types of cells associated with connective tissue.

a. _____

b. _____

c. _____

d. _____

D3.   Describe the characteristics and location of the following.

a. Hyaluronic acid

b. Chondroitin sulfate

c. Dermatan sulfate

d. Keratan sulfate

D4. Complete the following fill-in questions about connective tissue fibers.

a. _____ fibers consist of collagen plus some glycoprotein. They provide support and strength and also form the stroma of many soft organs.

b. _____ fibers are very tough and resistant to a pulling force, yet allow some flexibility in the tissue. They are composed of many minute fibers called fibrils.

c. _____ fibers are smaller than collagen fibers and freely branch and rejoin one another. These fibers also provide strength and can stretch up to 150% of their original length.

D5. After completing your study of connective tissue in the text, review Figure LG 3.1, G–L, and provide the following:

a. Names of the connective tissues.

b. One or more locations.

c. One or more functions.

D6. Check your understanding of embryonic and mature connective tissues by answering these questions.

a. *(Mesenchyme? Mucous connective tissue?)* is the tissue from which all other connective tissues derive.

b. The combining of areolar connective tissue with adipose tissue forms the

_____ layer.

c. The accumulation of a single large triglyceride droplet pushes the cytoplasm and nucleus to

the edge of the cell in this tissue: _____.

d. *(Dense regular? Dense irregular?)* connective tissue is associated with the heart valves, periosteum, joint capsules, and pericardium of the heart.

e. The cells of mature cartilage are called _____.

f. The surface of cartilage is surrounded by dense connective tissue called

_____.

D7. Match the following types of cartilage with their proper description or location.

E elastic cartilage	F fibrocartilage	H hyaline cartilage

a. _____ Maintains the shape of the epiglottis, external ear, and auditory tubes.

b. _____ Forms the intervertebral discs and pubic symphysis as well as the menisci of the knees.

c. _____ Found covering the ends of the long bones and anterior ends of the ribs, and helps to form parts of the nose, larynx, trachea, and bronchi.

D8. Fill in the correct terms pertaining to the osteon (Haversian system).

a. Each osteon consists of a _____ _____ that contains blood vessels and nerves.

b. _____ are concentric rings of matrix that consist of mineral salts.

c. The small spaces between lamellae are called _____.

d. Mature bone cells are referred to as _____.

e. Minute canals called _____ connect the lacunae and contain projections from the mature bone cells.

D9. Complete these questions about blood (vascular) tissue.

a. _____ is the straw-colored liquid that consists mostly of water with a wide variety of dissolved substances.

b. *(Erythrocytes? Platelets?)* function in transporting oxygen and carbon dioxide.

c. _____ function in blood clotting.

d. _____ are involved in phagocytosis, immunity, and allergic reactions.

E. Membranes   (pages 81–82)

E1. Test your knowledge of membranes by filling in the following table.

Name	Location	Functions
Mucous	a.	b.
c.	Lines body cavities not open directly to exterior	d.
e.	f.	Secretes fluid that lubricates articular cartilage at joints.

E2. Match each of the following serous membranes with its location.

> 1. pleura
> 2. pericardium
> 3. peritoneum

a. _____ Lines the heart cavity and covers the heart.

b. _____ Lines the thorax cavity and covers the lungs.

c. _____ Lines the abdominal cavity and covers the abdominal organs.

F. Muscle Tissue (pages 82–83)

F1. Match the description with the proper muscle tissue.

> 1. striated and voluntary
> 2. striated and involuntary
> 3. nonstriated and involuntary

a. _____ Smooth muscle   c. _____ Cardiac muscle

b. _____ Skeletal muscle

F2. After completing your study of muscle tissue in the text, review Figure LG 3.1, M–O, and complete these questions.

a. Name of the muscle tissue.

b. One or more locations of the tissues.

c. One or more functions of the tissues.

G. Nervous Tissue (page 83–84)

G1. Describe the following terms.

a. Cell Body

b. Dendrites

c. Axons

d. Neuroglia

G2. Refer to diagram P in Figure LG 3.1. Complete the following information.

a. Name of the tissue.

b. One or more locations.

c. One or more functions.

ANSWERS TO SELECT QUESTIONS

A1. Body, hollow, glands.
A2. Supports, protects, binds, energy, immunity.
A3. Movement.
A4. Transmits, potentials, coordinate.
B1.

Cell Junction	Functions
Tight	a. Prevent the passage of molecules between cells.
Plaque-bearing	b. Fasten cells to one another or to the extracellular material.
Gap	c. Allow the rapid spread of nerve impulses.

C1. Lining/covering, glandular.
C2. (a) F; (b) T; (c) T; (d) F; (e) T; (f) T.
C3. (a) P; (b) Si; (c) St.
C4.

Squamous	a. Flat
b. Cuboidal	Cube shaped
Columnar	c. Tall and cylindrical
d. Transitional	Varies in shape with distension

C5. (a) A – simple squamous; B – simple cuboidal; C – non-ciliated simple columnar; D – ciliated simple columnar; E – stratified squamous; F – pseudostratified.
C7. (a) Holocrine; (b) merocrine; (c) apocrine.
D1. (a) T; (b) F; (c) F; (d) T; (e) F.
D2. (a) Fibroblast; (b) macrophage; (c) plasma cell; (d) mast cell.
D4. (a) Reticular; (b) collagen; (c) elastic.
D5. (a) G – areolar connective tissue; H – adipose tissue; I – dense regular connective tissue; J – hyaline cartilage; K – bone (osseous tissue); L – blood (vascular tissue).

D6. (a) Mesenchyme; (b) subcutaneous; (c) adipocyte; (d) dense irregular; (e) chondrocytes; (f) perichondrium.
D7. (a) E; (b) F; (c) H.
D8. (a) Central canal; (b) lamellae; (c) lacunae; (d) osteocytes; (e) canaliculi.
D9. (a) Plasma; (b) erythrocytes; (c) platelets; (d) leukocytes.
E1.

Name	Location	Functions
Mucous	a. Lines body cavity that opens directly to the exterior	b. Produces mucus
c. Serosa	Lines body cavities not open directly to exterior	d. Secretes a lubricating fluid
e. Synovial	f. Freely movable joints	Secretes fluid that lubricates articular cartilage at joints.

E2. (a) 2; (b) 1; (c) 3.
F1. (a) 3; (b) 1; (c) 2.
F2. (a) M – skeletal muscle; N – cardiac muscle; O – smooth muscle.
G1. (a) Contains the nucleus and other organelles; (b) highly branched processes that conduct nerve impulses toward the cell body; (c) single long processes that conduct impulses away from the cell body; (d) cells that protect and support the neurons.
G2. (a) P – nervous tissue

SELF QUIZ

Choose the one best answer to the following questions.

1. Which of the following is specialized for contractility?

   A. epithelial tissue
   B. loose connective tissue
   C. nervous tissue
   D. muscle tissue
   E. adipose tissue

2. The cell junction that prevents movement of a substance through intercellular routes is a(n)

   A. gap junction
   B. tight junction
   C. adherens junction
   D. A and B are both correct
   E. B and C are both correct

3. _____ epithelium lines part of male urethra, large excretory ducts of some glands, and a small area of anal mucous membranes.

A. stratified squamous
B. simple columnar
C. simple cuboidal
D. stratified columnar
E. pseudostratified

4. The modifications to the columnar epithelium include

A. microvilli
B. cilia
C. mucus production (goblet cells)
D. all of the above
E. A and B only

5. Basement membranes are characteristically associated with which of the following tissues?

A. hyaline cartilage
B. muscle
C. pseudostratified columnar
D. nervous
E. osseous

6. The secretion from which type of gland involves the death and discharge of the cell producing the secretion?

A. apocrine
B. holocrine
C. merocrine
D. A and B are correct
E. B and C are correct

7. Which of the following is NOT associated with the osteon?

A. lacuna
B. canaliculi
C. lamella
D. chondrocyte
E. osteocyte

8. Which of the following tissues is avascular?

A. bone
B. cartilage
C. epithelium
D. loose connective
E. B and C

9. Which type of membrane does NOT contain epithelium?

A. mucous
B. serous
C. cutaneous
D. synovial
E. A and B are correct

10. _____ muscle tissue is characterized by branched cylinder-shaped cells, only one nucleus, and intercalated discs which contain anchoring and communicating junctions.

A. skeletal
B. smooth
C. cardiac
D. A and B are both correct
E. none of the above are correct

11. Which type of connective tissue forms tendons and ligaments?

A. areolar
B. dense regular
C. dense irregular
D. elastic
E. reticular

12. The protein substance in epithelium tissue that is resistant to friction is called

A. hyaluronic acid
B. chondroitin sulfate
C. keratin
D. dermatan sulfate
E. keratan sulfate

13. The term matrix refers to _____, which are outside the cells.

A. intercalated discs
B. ground substance and fibers
C. collagen fibers only
D. elastic fibers only
E. reticular fibers only

14. _____ are single, long processes of the neuron that conduct nerve impulses away from the cell body.

A. axons
B. dendrites
C. neuroglia
D. none of the above are correct

15. The type of tissue that lines the bladder is

A. simple columnar
B. transitional
C. dense irregular
D. areolar
E. pseudostratified

Answer (T) True or (F) False to the following questions.

16. _____ Cardiac muscle is striated and voluntary.

17. _____ Fibroblasts are large, flat cells with branching processes that secrete matrix of connective tissue.

18. _____ Hyaluronic acid binds cells together and maintains the shape of the eyeballs.

19. _____ The major antibody-producing cells of the body are mast cells.

20. _____ Reticular fibers provide support and strength and also form the stroma of many soft organs.

Fill in the blanks.

21. All connective tissue is derived from _____.

22. _____ surrounds the surface of most cartilage.

23. _____ glands have ducts and secrete their products to the surface through these ducts.

24. The word that would best describe the blood supply to all connective tissue except cartilage is

_____.

25. The serous membrane that lines the thoracic cavity and covers the lungs is called the _____.

ANSWERS TO THE SELF QUIZ

1. D	10. C	19. F
2. B	11. B	20. T
3. D	12. C	21. Mesenchyme
4. D	13. B	22. Perichondrium
5. C	14. A	23. Exocrine
6. B	15. B	24. Vascular
7. D	16. F	25. Pleura
8. E	17. T	
9. D	18. T	

The Integumentary System

SYNOPSIS

You have now completed the first leg of your journey and it is time for you to continue on your "fantastic voyage." This chapter will introduce you to the first system on your itinerary.

A group of tissues performing a specific function constitutes an **organ**, while a **system** consists of a group of organs functioning together to perform a specific task.

The **integumentary system** includes the **skin**, **hair**, **nails**, **glands**, and several specialized nervous system receptors. Though often overlooked, the integumentary system serves as the largest interactive system between us and the environment. It provides us with protection from disease, ultraviolet light, and injury, as well as functioning in excretion, thermoregulation, reception of stimuli, and the synthesis of vitamin D.

The skin is one of our largest organs, having a surface area of approximately 2 square meters. Microscopically thin, it has a complex organization which is easily injured and yet possesses the capacity to repair or replace the damaged and worn segments.

TOPIC OUTLINE AND OBJECTIVES

A. Skin

1. Define the integumentary system.
2. List and describe the components of the epidermal layer, detailing their structure and function.
3. Describe the structural and functional components of the dermis.
4. Discuss the origin of skin pigmentation and its function.
5. Describe epidermal ridges.
6. Discuss the blood supply to the skin.

B. Epidermal Derivatives

7. Contrast the structure, function, and location of hair, nails, and skin glands.

C. Developmental Anatomy of the Integumentary System

D. Aging and the Integumentary System

E. Applications to Health

8. Discuss the cause and effects of burns, sunburns, skin cancer, acne, and pressure sores.

F. Key Medical Terms Associated with the Integumentary System

SCIENTIFIC TERMINOLOGY

Find an anatomical sample word for each prefix and suffix:

Prefix/Suffix	Meaning	Sample Word
ab-	away	
angio-	blood	
cera-	wax	
dermato-	skin	
endo-	within	
ecto-	outside	
epi-	above	
erythros-	red	
-ferre	to bear	
germ-	sprout	
hemo-	blood	
jaune-	yellow	
kyanos-	blue	
intra-	within	
melan-	black	
-oma	tumor	
onyx-	nail	
sebo-	grease	
sudor-	sweat	

A. Skin (pages 91–97)

A1. Refer to Figure LG 4.1.

a. Using colored pencils, shade in the five layers associated with the epidermis (be sure that your coloring corresponds with the color code ovals).

   O stratum basale

   O stratum corneum

   O stratum granulosum

   O stratum lucidum

   O stratum spinosum

b. Label the following structures.
   1. corpuscle of touch
   2. dermal papilla
   3. excretory duct
   4. lamellated corpuscle
   5. reticular region
   6. papillary region
   7. sudoriferous gland

c. Label the five layers of the epidermis.

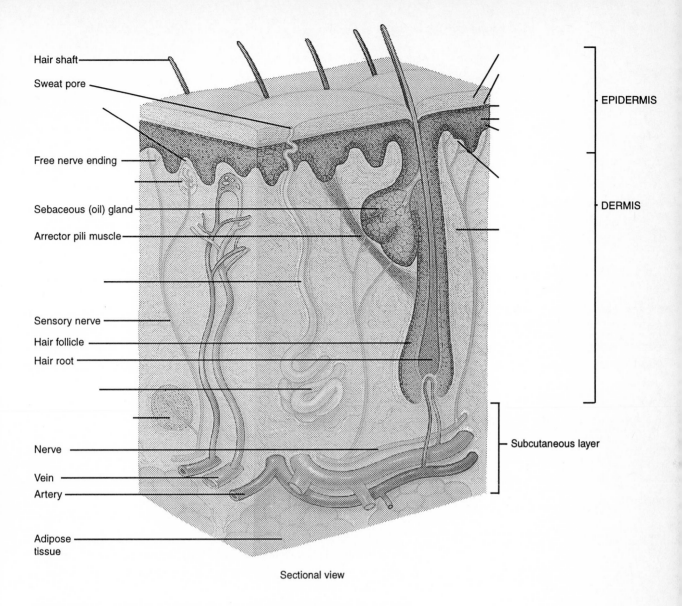

Hair shaft

Sweat pore

Free nerve ending

Sebaceous (oil) gland

Arrector pili muscle

Sensory nerve

Hair follicle

Hair root

Nerve

Vein

Artery

Adipose
tissue

EPIDERMIS

DERMIS

Subcutaneous layer

Sectional view

Figure LG 4.1   Structure of the skin and underlying subcutaneous tissue.

A2.   Match the following skin layers with their description.

> D   dermis
> E   epidermis
> H   hypodermis

a.   _____ The deepest layer, also called the superficial fascia, consists of areolar and adipose
tissues.

b.   _____ The superficial, thinner portion, which is composed of epithelium.

c.   _____ The inner, thicker connective tissue layer.

A3. List and give the functions of the four distinct types of cells found in the epidermis.

a.

b.

c.

d.

A4. Match the following epidermal layers to their description.

1. stratum basale	4. stratum lucidum
2. stratum corneum	5. stratum spinosum
3. stratum granulosum	

a. _____ This layer of the epidermis contains 8–10 rows of polyhedral keratinocytes that fit closely together.

b. _____ This layer consists of 25–30 rows of flat, dead keratinocytes completely filled with keratin.

c. _____ Found only in the thick skin of the palms and soles and is absent in thin skin.

d. _____ Consists of 3–5 rows of flattened keratinocytes that contain darkly staining granules of a substance called keratohyalin.

e. _____ This single layer of cuboidal to columnar keratinocytes contains stem cells, which are capable of continued cell division. Melanocytes, Langerhans cells, and Merkel cells are scattered here.

A5. What is keratin and what is its function?

A6. Dermis; answer (T) true or (F) false to the following questions.

a. _____ The cells associated with the dermis include the fibroblast, macrophage, and chondrocyte.

b. _____ The dermis is very thick on the palms and soles.

c. _____ There are relatively few blood vessels, nerves, and glands embedded in the dermis.

d. _____ The outer region of the dermis is called the papillary region.

e. _____ Corpuscles of touch are often found within the dermal papillae.

f. _____ The reticular layer consists of dense, irregular connective tissue containing collagen and elastic fibers.

g. _____ Lamellated corpuscles are sensitive to light touch.

h. _____ Striae are the red lines that occur with stretching.

A7. List the seven functions associated with the skin.

a. _____

b. _____

c. _____

d. _____

e. _____

f. _____

g. _____

A8. Skin Color; fill in the blanks.

a. The partial or complete loss of melanocytes from areas of skin, producing irregular white

   spots, is called _____.

b. An inherited inability of an individual in any race to produce melanin results in

   _____.

c. A pigment in the epidermis, _____ , is synthesized by _____ ,
   and varies the skin color from pale yellow to black.

d. _____ , a yellow-orange pigment, is found in the stratum corneum and fatty
   areas of the dermis and subcutaneous layer in people of Asian ancestry.

e. The pink to red color of Caucasian skin is due to _____.

f. The most lethal skin cancer in young women is _____

   _____.

A9. Define epidermal ridges.

B. Epidermal Derivatives (pages 97–100)

B1. What is the primary function of hair?

B2. What factors affect the rate of hair loss?

B3. Complete the following questions pertaining to the anatomy of the hair.

a. Surrounding the root is the _____, which is made up of an external root sheath and internal root sheath of epithelium.

b. The _____ of the hair is the outermost layer, consisting of a single layer of thin, flat, scalelike cells.

c. The _____ is the superficial portion, which projects from the surface of the skin.

d. Composed of two or three rows of polyhedral cells containing pigment granules and air

spaces, this portion is the _____.

e. The _____ forms the major part of the shaft and consists of elongated cells, which contain pigment granules in dark hair.

f. The onion-shaped structure at the base of each follicle is the _____.

g. The bulb contains a ring of cells called the matrix, which is the _____ layer.

h. The _____ contains many blood vessels and provides nourishment for the growing hair.

i. The contraction of the _____ _____ result in "goose-bumps" or "gooseflesh."

j. Around each hair follicle are nerve endings called _____

_____ _____.

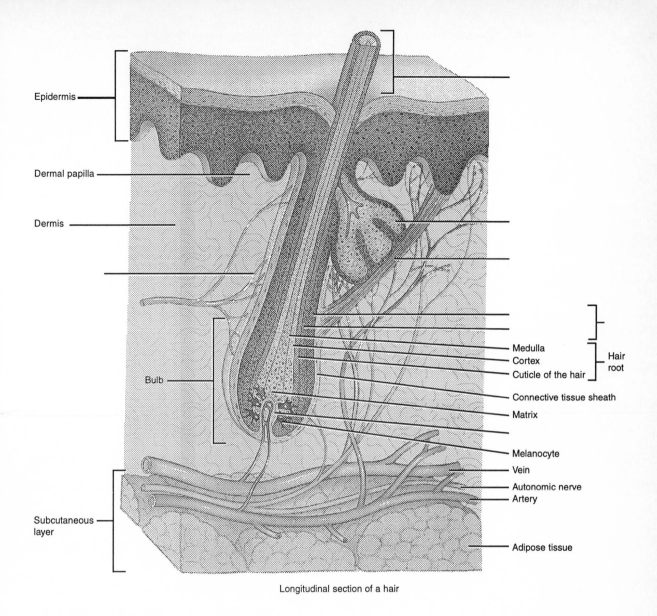

Epidermis

Dermal papilla

Dermis

Bulb

Subcutaneous
layer

Medulla
Cortex
Cuticle of the hair

Hair
root

Connective tissue sheath
Matrix

Melanocyte
Vein
Autonomic nerve
Artery

Adipose tissue

Longitudinal section of a hair

Figure LG 4.2   Longitudinal section of a hair.

B4.   Refer to Figure LG 4.2 and follow the directions below.

a.   Label the following structures.
- ○   arrector pili
- ○   hair follicle
- ○   hair root plexus
- ○   hair shaft
- ○   papilla of hair
- ○   sebaceous gland

b.   Using colored pencils, shade the structures listed above. Make sure your colors correspond
     to the color code ovals.

B5. Check your understanding of skin glands. You may use an answer more than once.

A apocrine sweat glands	SE sebaceous
C ceruminous	SU sudoriferous
E eccrine sweat glands	

a. _____ These glands are distributed in the axilla, pubic region, and pigmented areas of the breast.

b. _____ The scientific name for a sweat gland.

c. _____ With few exceptions, these glands are associated with hair follicles.

d. _____ These are modified sudoriferous glands present in the external auditory canal.

e. _____ Secrete an oily substance.

f. _____ They are distributed throughout the skin except for such places as the margins of the lips, nail beds, glans penis, and eardrum.

g. _____ The enlargement of this gland leads to the formation of blackheads, pimples, and boils.

B6. Which terms correspond to the following descriptions?

cuticle	nail body
free edge	nail matrix
lunula	nail root

a. The _____ is a narrow band of epidermis that extends from the margin of the nail wall, adhering to it.

b. The whitish semilunar area of the proximal end of the body is called the

_____.

c. The _____ is the portion of the nail that is visible.

d. The part that may extend beyond the distal end of the digit is the _____.

e. The _____ is the portion buried in a fold of skin.

f. The epithelium, deep to the nail root, is known as the _____.

B7. Differentiate between

a. cutaneous plexus

b. papillary plexus

C. Developmental Anatomy of the Integumentary System (pages 100–101)

C1. List the week that the following structures begin development.

a. Nails _____

b. Hair follicles _____

c. Epidermis _____

D. Aging and the Integumentary System   **(page 102)**

D1.   List several changes associated with the aging of the integument.

E. **Applications to Health** (Refer to *Applications to Health* companion to answer Part
   E questions.)

E1.   List the five systemic effects of a burn.

a.

b.

c.

d.

e.

E2.   List the risk factors for skin cancer.

a.

b.

c.

d.

e.

E3.   Match the following clinical disorders to their description.

| A | acne |
| D | decubitus ulcers |

a.   _____ Caused by a constant deficiency of blood to tissues overlying a bony projection
   that has been subjected to prolonged pressure.
b.   _____ An inflammation of sebaceous glands that usually begins at puberty. Sebaceous
   follicles are colonized by bacteria.

E4.   Explain how overexposure to ultraviolet (UV) light (sunburn) leads to cell dysfunction
   and death.

ANSWERS TO SELECT QUESTIONS

A2. (a) H; (b) E; (c) D.

A3. (a) Keratinocytes—produce the protein keratin; (b) melanocytes—produce melanin (pigment); (c) Langerhans cells—interact with helper T cells (immunity); (d) Merkel cells—are thought to function in the sensation of touch.

A4. (a) 5; (b) 2; (c) 4; (d) 3; (e) 1.

A5. Protein that protects skin from abrasion, microbes, heat waves, and many chemicals.

A6. (a) F; (b) T; (c) F; (d) T; (e) T; (f) T; (g) F; (h) F.

A7. (a) Regulates body temperature; (b) protects; (c) sensation; (d) excretion; (e) immunity; (f) blood reservoir; (g) synthesis of vitamin D.

A8. (a) Vitiligo; (b) albinism; (c) melanin, melanocytes; (d) carotene; (e) hemoglobin; (f) malignant melanoma.

B1. Protection.

B2. Illness, diet, high fever, surgery, blood loss, stress.

B3. (a) Hair follicle; (b) cuticle; (c) shaft; (d) medulla; (e) cortex; (f) bulb; (g) germinal; (h) papilla; (i) arrector pili; (j) hair root plexuses.

B5. (a) A; (b) SU; (c) SE; (d) C; (e) SE; (f) E; (g) SE.

B6. (a) Cuticle; (b) lunula; (c) nail body; (d) free edge; (e) nail root; (f) nail matrix.

C1. (a) Tenth; (b) ninth–twelfth; (c) eighth.

E1. (a) Loss of water, plasma, and plasma proteins; (b) bacterial infection; (c) reduced blood circulation; (d) decreased urine production; (e) diminished immune response.

E2. (a) Skin type; (b) sun exposure; (c) age; (d) family history; (e) immunological status.

E3. (a) D; (b) A.

SELF QUIZ

Choose the one best answer to the following questions.

1. Goose bumps are a result of

   A. contraction of the dermal papillae
   B. contraction of elastic fibers in the hair follicle
   C. conscious action
   D. contraction of the arrector pili
   E. secretion of perspiration

2. In the joining of the epidermis and dermis, which two layers are involved?

   A. basale and reticular
   B. germinativum and lucidum
   C. basale and subcutaneous
   D. basale and papillary
   E. spinosum and reticular

3. Which of the following is properly matched?

   A. sudoriferous—oil
   B. sebaceous—sweat
   C. sudoriferous—keratin
   D. sebaceous—oil
   E. none are properly matched

4. If you were dissecting the palm of a hand, the third layer of the epidermis you would cut is the stratum

   A. spinosum
   B. granulosum
   C. corneum
   D. lucidum
   E. basale

5. The stratum corneum consists of

   A. mitotically active cells
   B. cells that produce melanin
   C. dead cells
   D. cells containing adipose tissue
   E. fiber-producing cells

6. Acne is an inflammation of the

   A. oil glands
   B. sweat glands
   C. epithelium
   D. ceruminous glands
   E. sudoriferous glands

7. Merkel (tactile) discs are located in the stratum
   _____.

   A. basale
   B. corneum
   C. granulosum
   D. germinativum
   E. lucidum

8. Which portion of a nail is buried in a fold of skin?

   A. nail body
   B. nail root
   C. lunula
   D. eponychium
   E. cuticle

9. Which gland begins to function at puberty and produces a viscous secretion?

A. sebaceous
B. apocrine sudoriferous
C. ceruminous
D. eccrine sudoriferous

10. Which portion of hair is responsible for providing nourishment for growth?

A. cuticle
B. papilla
C. cortex
D. follicle
E. medulla

11. _____ accounts for over 75% of all skin cancers.

A. solar keratosis
B. basal cell carcinoma
C. squamous cell carcinoma
D. malignant melanoma
E. none of the above are correct

Arrange the answers in the correct order.

12. From most superficial to deepest: _____ _____ _____
A. dermis
B. epidermis
C. superficial fascia

Answer (T) True or (F) False to the following questions.

13. _____ One of the functions of the integument is the production of vitamin B.

14. _____ The number of melanocytes is approximately equal in all races.

15. _____ The differences in skin coloration are related to the quantity of keratin in the skin.

16. _____ The most lethal skin cancer in young women is malignant melanoma.

17. _____ The loss of melanocytes from an area of skin, which produces irregular, white spots, is called albinism.

Fill in the blanks.

18. The _____ glands line the external auditory canal.

19. The dermis has two regions: the _____ region is the more superficial.

20. The _____ layer is the second most superficial layer of thick skin.

21. Pink to red coloration of Caucasian skin is due to _____, a red pigment in red blood cells.

22. The regulation of body temperature is directly related to the secretion of the _____ glands.

23. _____ is a localized tumor of the skin and subcutaneous layer that results from an abnormal increase in blood vessels.

24. _____ refers to a condition of the skin marked by reddened elevated patches that are often itchy; it is also called urticaria.

25. The nerve ending located in the subcutaneous layer is the _____ _____.

ANSWERS TO THE SELF QUIZ

1. D
2. D
3. D
4. B
5. C
6. A
7. A
8. B
9. B

10. B
11. B
12. B, A, C
13. F
14. T
15. F
16. T
17. F
18. Ceruminous

19. Papillary
20. Lucidum
21. Hemoglobin
22. Sudoriferous
23. Hemangioma
24. Hives
25. Lamellated (Pacinian) corpuscle

Bone Tissue

SYNOPSIS

If I were to throw out the unfinished phrase, "Dry as," you would probably finish with the words, "a bone." **Bone tissue** is, more often than not, thought of as a dry, inanimate substance found in the body. Though this concept is perpetuated by thousands of students examining the long-dead skeletal remains hanging in the corner of some biology laboratory, nothing could be further from the truth.

Bone or **osseous tissue** is a highly complex, extremely vascular, living ("breathing") tissue, composed of millions of cells. Bone tissue possesses the ability to react to stress by re-modeling itself: it provides support and protection, allows for movement (via skeletal muscle contraction), serves in the storage of minerals and energy, and is the site of blood cell production.

Not completely solid, bone **exhibits** numerous spaces that provide passageways for blood vessels and nerves. These spaces also make bone lighter. Strength plus lightness is the key to bone's ability to perform its many functions.

Your journey continues as this chapter takes you into the macroscopic and microscopic worlds of bone tissue. You will venture through nutrient and vascular channels into the very essence of bone tissue as you view its many structural wonders.

TOPIC OUTLINE AND OBJECTIVES

A. Functions of Bone

1. List and explain the six functions of bone tissue.

B. Anatomy and Histology of Bone

2. Discuss the chemical nature of bone tissue.
3. List and describe the four types of cells associated with bone tissue.
4. List and describe the seven parts of a long bone.
5. Differentiate between compact and spongy bone tissue.

C. Bone Formation: Ossification

6. Contrast intramembranous and endochondral ossification.

D. Bone Growth and Replacement

7. Explore the processes of bone growth and replacement.
8. Describe the mechanism of bone remodeling.

E. Blood and Nerve Supply

F. Exercise and Bone Tissue

G. Developmental Anatomy of the Skeletal System

H. Aging and Bone Tissue

I. Applications to Health

9. Define osteoporosis, Paget's disease, and fractures.
10. Differentiate among several common fractures.

J. Key Medical Terms Associated with Bone Tissue

SCIENTIFIC TERMINOLOGY

Find an anatomical sample word for each prefix and suffix:

Prefix/Suffix	Meaning	Sample Word
arthro-	joint	
chondro-	cartilage	
-clast	to blast	
-cyte	cell	
dia-	through	
endo-	within	
-itis	inflammation	
medulla-	central part of an organ	
myelos-	marrow	
osteo-	bone	
-penia	poverty	
-physis	growth	
pro-	precursor	

A. Functions of Bone   (page 108)

A1.   List the six functions of bone.

a. _____

b. _____

c. _____

d. _____

e. _____

f. _____

B. Anatomy and Histology of Bone   (pages 108–113)

B1.   Complete the following questions pertaining to bone histology.

a. The matrix of bone is about _____% mineral salts, _____% protein fibers, and

 _____% water.

b. The cells that possess the ability to differentiate into osteoblasts are called

 _____.

c. _____ or mature bone cells are the principal cells of bone tissue.

d. The cells that are associated with bone formation are _____.

e. _____ develop from circulating monocytes and function in bone removal by resorption.

f. Calcium _____ and tricalcium _____ compose much of the mineral salts found in bone.

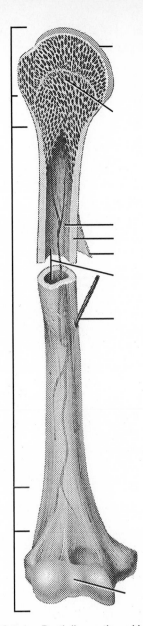

Figure LG 5.1 Partially sectioned long bone.

B2. Label the diagram, Figure LG 5.1, of a typical long bone, using the following terms.

proximal epiphysis   articular cartilage
diaphysis   spongy bone
distal epiphysis   compact bone
endosteum   medullary cavity
periosteum

B3. Contrast compact (dense) and spongy (cancellous) bone tissue.

a. Compact bone tissue

b. Spongy bone tissue

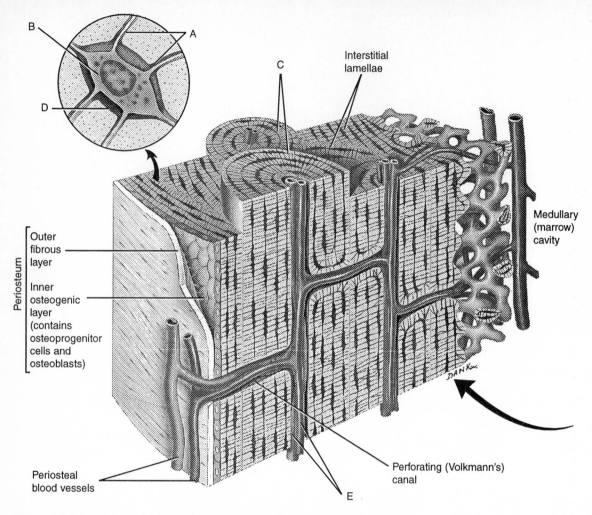

Figure LG 5.2 Enlarged aspect of several osteons in compact bone.

B4. Label Figure LG 5.2 and then answer the following questions about the osteon.

a. After entering the bone through perforating canals, the blood vessels and nerves run

longitudinally through the bone in _____ _____. On

the diagram it is labeled _____.

b. The compact bone which radiates in rings around the central canal is referred to as

_____ _____. It is labeled _____.

c. Between these concentric rings of compact bone are small spaces called

_____. Label with letter _____.

d. Mature bone cells known as _____ are located in the small spaces or

"little lakes." It is labeled _____.

e. The minute channels that connect the small spaces of the osteon are called

_____. Label with letter _____.

B5. Check your understanding of spongy bone.

a. The irregular latticework of thin plates of bone is called _____.

b. What substance fills the spaces within spongy bone? _____.

C. Bone Formation: Ossification   (pages 113–116)

C1. Check your understanding of bone formation.

a. The process of bone formation is called _____.

b. The formation of bone directly on or within fibrous connective tissue membranes is

called _____ _____.

c. _____ _____ refers to the formation of bone in hyaline cartilage.

C2. Briefly list the five phases of endochondral ossification.

D. Bone Growth and Replacement   (pages 116–117)

D1. Answer the questions below using the terms in the box.

calcified matrix	proliferating cartilage
epiphyseal line	resting cartilage
hypertrophic cartilage	

a. The appearance of the _____ _____ indicates that bone growth in length has stopped.

b. The zone of _____ _____ consists of slightly larger chondrocytes arranged like stacks of coins.

c. The lengthwise expansion of the epiphyseal plate is the result of cell divisions at the zone of

proliferating cartilage and maturation of cells in the zone of _____

_____.

d. The zone of _____ _____ is only a few cells thick and consists mostly of dead cells.

e. The cells that act to anchor the epiphyseal plate to the bone of the epiphysis are in the

zone of _____ _____.

D2. Answer the following questions relating to bone replacement.

a. Which of the following minerals are important to normal bone growth? (Circle all correct answers.)

boron   manganese

calcium   phosphorus

iodine   selenium

magnesium   zinc

b. Which vitamins are important to bone growth and replacement? (Circle all the correct answers.)

Vit. A   Vit. B_{12}

Vit. B_1 Vit. C

Vit. B_2 Vit. D

Vit. B_6 Vit. E

c. True or false: Bone is continually replacing itself throughout adult life.

D3. List the hormones that are essential for proper bone activity.

a. _____

b. _____

c. _____

d. _____

e. _____

f. _____

E. Blood and Nerve Supply (page 117)

E1. Define the terms listed below.

a. Nutrient artery

b. Nutrient foramen

c. Epiphyseal arteries

d. Periosteal arteries

E2. Discuss the nerve supply to bone.

F. Exercise and Bone Tissue (page 120)

F1. Describe the result of lack of stress on bone tissue, and list two situations in which this would occur.

G. Developmental Anatomy of the Skeletal System (page 123)

G1. Give the date (in weeks) that the following events occur.

a. Limb buds _____

b. Hand and foot plates _____

c. Evidence of the arm and forearm _____

d. Evidence of the shoulder _____

H. Aging and Bone Tissue (page 123)

H1. Describe the two principal effects of aging on the skeletal system.

a.

b.

I. Applications to Health (Refer to *Applications to Health* companion to answer Part I questions.)

I1. Define osteoporosis and list the 9 factors implicated in it.

a. Osteoporosis

b.

1. _____

2. _____

3. _____

4. _____

5. _____

6. _____

7. _____

8. _____

9. _____

I2. Briefly describe Paget's disease.

I3. Answer the questions below using the following terms.

1. closed	4. partial
2. complete	5. pathological
3. open	6. stress

a. _____ A fracture due to weakening of a bone caused by a disease process such as osteoporosis.

b. _____ A fracture in which the break across the bone is incomplete.

c. _____ A fracture resulting from training on hard surfaces, or longer distances, or greater speed.

d. _____ A fracture in which the bone is broken into two or more pieces.

e. _____ A fracture in which the broken end protrudes through the skin.

f. _____ A fracture in which the broken end does not protrude through the skin.

A1. (a) support; (b) protection; (c) movement;
(d) mineral storage; (e) blood cell production;
(f) storage of energy.

ANSWERS TO SELECTED QUESTIONS

B1. (a) 50, 25, 25; (b) osteoprogenitors; (c) osteocytes; (d) osteoblasts; (e) osteoclasts; (f) carbonate, phosphate.

B3. (a) Contains few spaces, covers spongy bone, is thicker in the diaphysis; (b) contains many spaces filled with red bone marrow.

B4. (a) central canals, E; (b) concentric lamellae, C; (c) lacunae, D; (d) osteocytes, B; (e) canaliculi, A.

B5. (a) Trabeculae; (b) red bone marrow.

C1. (a) ossification or osteogenesis; (b) intramembranous ossification; (c) endochondral ossification.

D1. (a) epiphyseal line; (b) proliferating cartilage; (c) hypertrophic cartilage; (d) calcified matrix; (e) resting cartilage.

D2. (a) Boron, calcium, magnesium, manganese, phosphorus; (b) vitamins D, C, A, B_{12}; (c) true.

D3. (a) Human growth hormone; (b) calcitonin; (c) parathyroid hormone; (d) estrogen/testosterone; (e) insulin; (f) thyroid hormone.

F1. Loss of bone mass (decalcification) occurs in casting a broken limb due to reduced mechanical stress and in astronauts due to weightlessness.

G1. (a) Fifth; (b) sixth; (c) seventh; (d) eighth.

H1. (a) loss of calcium from the bone; (b) decreased rate of protein formation, therefore decreased rate of matrix production.

I3. (a) 5; (b) 4; (c) 6; (d) 2; (e) 3; (f) 1.

SELF QUIZ

Choose the one best answer to the following questions.

1. Which of the following is NOT a function of the skeletal system?

   A. support
   B. protection
   C. movement
   D. excretion
   E. storage of energy

2. Vitamin _____ is essential for bone growth and development.

   A. B$_1$
   B. E
   C. D
   D. B$_{13}$
   E. B$_6$

3. The zone of _____ consists of slightly larger chondrocytes arranged like stacks of coins.

   A. calcified matrix
   B. hypertrophic cartilage
   C. proliferating cartilage
   D. resting cartilage
   E. epiphyseal plate

4. Which of the following hormones is NOT involved in bone tissue activity?

   A. human growth hormone
   B. parathyroid hormone
   C. calcitonin
   D. estrogen and testosterone
   E. all are involved in bone tissue activity

5. These cells are associated with bone formation but have lost their ability to divide by mitosis.

   A. osteocytes
   B. osteoblasts
   C. osteoclasts
   D. osteons
   E. osteoprogenitors

6. The shaft or main portion of a long bone is called the

   A. epiphysis
   B. metaphysis
   C. endosteum
   D. diaphysis
   E. periosteum

7. Which of the following cells have the ability to undergo mitosis and develop into osteoblasts?

   A. osteoprogenitor
   B. osteocyte
   C. osteoclast
   D. osteoblast
   E. B and C are correct answers

8. The primary organic constituent of bone tissue is

   A. tricalcium phosphate
   B. collagen
   C. keratin
   D. calcium phosphate
   E. sodium phosphate

9. Osteomyelitis is

   A. an inflammation of a bone
   B. a type of bone cancer
   C. a malignant tumor composed of bone tissue
   D. often caused by *Staphylococcus aureus*
   E. both A and D

10. Ossification begins around the _____ week of embryonic life.

    A. fifth to sixth week
    B. sixth to seventh week
    C. seventh to eighth week
    D. eighth to ninth week
    E. tenth to eleventh week

Answer (T) True or (F) False to the following questions.

11. _____ Osteoclasts are bone resorptive cells.

12. _____ Endochondral ossification describes bone developing within connective tissue membranes.

13. _____ All bone tissue when first formed is of the spongy type.

14. _____ The structure located between the diaphysis and epiphysis, which allows for elongation of the diaphysis, is the epiphyseal plate.

15. _____ Compact bone is chemically different from spongy bone.

16. _____ The perforating (Volkmann's) canal is the same as the central canal.

17. _____ A stress fracture could occur in a soldier's foot after a 30-mile hike.

18. _____ An osteosarcoma is a malignant tumor composed of bone tissue.

19. _____ Remodeling refers to the destruction of old bones by osteoblasts, while new bone is constructed by osteoclasts.

20. _____ The artery to the epiphysis of a long bone is referred to as the nutrient artery.

Fill in the blanks.

21. Blood cell formation, known as _____, occurs in the red bone marrow.

22. Yellow bone marrow is composed primarily of _____ cells.

23. Mineral salts compose about _____% of the weight of bone.

24. _____ bone tissue makes up most of the bone tissue of short, flat, and irregularly shaped bones and most of the epiphyses of long bones.

25. Actively growing connective tissue organized by capillary infiltration into a fracture homatoma is called a

_____.

ANSWERS TO THE SELF QUIZ

1. D	10. B	19. F
2. C	11. T	20. F
3. C	12. F	21. Hemopoiesis
4. E	13. T	22. Adipose
5. B	14. T	23. 50
6. D	15. F	24. Spongy
7. A	16. F	25. Procallus
8. A	17. T	
9. E	18. T	

The Skeletal System: The Axial Skeleton

CHAPTER

6

SYNOPSIS

When a skyscraper is under construction the workers first clear the land and then sink foundations deep into the earth. Next they begin to assemble a steel framework reaching into the sky. It is upon this framework that the bricks and mortar, metal, and glass are affixed.

The **skeletal system** forms the framework for our body. It supports the soft tissues, provides points of attachment for many skeletal muscles, and contributes to movement such as walking, throwing, and breathing. The bones of our body also serve as anatomical landmarks for many health practitioners.

The adult human skeleton consists of 206 named bones organized into two principal divisions: the **axial skeleton** and the **appendicular skeleton.** Our journey continues as we explore the 80 bones that comprise the axial skeleton.

TOPIC OUTLINE AND OBJECTIVES

A. Divisions of the Skeletal System

1. Contrast the components of the axial and appendicular skeletal systems.

B. Types of Bones

2. List and describe the physical features of the four types of bones.

C. Bone Surface Markings

3. List, describe, and give examples of 15 bone surface markings.

D. Skull

4. List and identify the 22 bones of the skull.
5. Define and identify the principal sutures and fontanels.
6. Define, identify, and list the four pairs of paranasal sinuses and the 22 pairs of foramina of the skull.

E. Hyoid Bone

F. Vertebral Column

7. Identify the bones of the vertebral column and their principal characteristics.
8. List the defining features and normal curves of each anatomical region of the vertebral column.

G. Thorax

9. Identify the bones of the thorax.
10. List the defining features of the sternum and ribs.

H. Applications to Health

11. Contrast the disorders of a herniated disc, scoliosis, kyphosis, lordosis, spina bifida, and fractures of the vertebral column.

SCIENTIFIC TERMINOLOGY

Find an anatomical sample word for each prefix and suffix:

Prefix/Suffix	Meaning	Sample Word
corona-	crown	
costa-	rib	
epi-	above	
-glossus	tongue	
infra-	below	
inter-	between	
kyphos-	hunchback	
lordo-	swayback	
para-	beside	
paries-	wall	
sagitta-	arrow	
scolio-	bent	
spheno-	wedge shaped	
squama-	flat	
stylo-	stake or pole	
supra-	above	
sutura-	seam	
tempora-	temples	
tube-	knob	
xipho-	sword-like	

A. Divisions of the Skeletal System   (page 126)

A1. There are _____ bones in the axial division and _____ bones in the appendicular division.

A2. The bones called _____ connect the limbs to the axial skeleton.

B. Types of Bones   (page 126–128)

B1. Using the six terms below, match the description to the bone type (you may use an answer more than once).

1. flat bones	4. sesamoid bones
2. irregular bones	5. short bones
3. long bones	6. sutural bones

a. _____ Small bones located between the joints of certain cranial bones.

b. _____ Have a greater length than width, and consist of a diaphysis and a variable number of extremities.

c. _____ Composed of two nearly parallel plates of compact bone enclosing a layer of spongy bone.

d. _____ Somewhat cube-shaped and nearly equal in length and width.

e. _____ Have complex shapes; include the vertebrae and certain facial bones.

f. _____ These are small bones in tendons where considerable pressure develops.

g. _____ Bones of the thighs, legs, toes, arms, forearms, and fingers are examples of this type of bone.

h. _____ Cranial bones, sternum, ribs, and scapulae are classified as this type of bone.

i. _____ The patella (kneecap) would be classified this way.

C. Bone Surface Markings   (pages 128–129)

C1.   List several functions of surface markings.

C2.   Complete the following table.

Marking	Description	Example
Fissure		
	Opening through which blood vessels, nerves, and ligaments pass	
Sulcus		
	A depression in or on a bone	
Facet		
	A large, rounded, usually roughed process	
Trochanter		
Crest		
Spinous process		
	A prominence above a condyle	

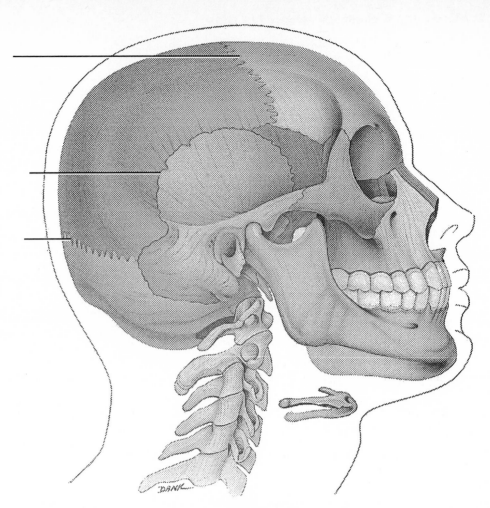

Right lateral view

Figure LG6.1   Skull.

D. Skull   (pages 129–149)

D1. On Figure LG 6.1, label the coronal, squamous, and lambdoid sutures.

D2. Color the skull bones on Figure LG 6.1. Be sure the color corresponds to the color code oval for each bone listed on the figure.

- ⭕ ethmoid bone
- ⭕ frontal bone
- ⭕ lacrimal bone
- ⭕ mandible bone
- ⭕ maxilla bone
- ⭕ nasal bone
- ⭕ occipital bone
- ⭕ sphenoid bone
- ⭕ temporal bone
- ⭕ zygomatic bone

D3. Check your knowledge of the skull by filling in the blanks for the questions below.

a. The skull contains _____ bones.

b. The _____ bones enclose and _____ the brain.

c. There are _____ facial bones in the skull.

d. A _____ is an immovable joint found only between skull bones.

e. At birth, membrane-filled spaces called _____ are found between cranial bones.

D4. Match the following sutures or fontanels with the correct description (refer to Table 6.5, and Figures 6.4, 6.6, and 6.14 in the textbook).

1. anterior	5. posterior
2. anterolateral	6. posterolateral
3. coronal	7. sagittal
4. lambdoid	8. squamous

a. _____ These fontanels are located on each side of the skull at the junction of the frontal, parietal, temporal, and sphenoid bones.

b. _____ The suture between the parietal bones and occipital bone.

c. _____ This fontanel is situated between the two parietal bones and the occipital bones.

d. _____ The suture located between the frontal bone and the parietal bones.

e. _____ This suture is located between the two parietal bones.

f. _____ The fontanel located between the angles of the two parietal bones and the two segments of the frontal bones.

g. _____ The suture between the parietal bones and temporal bones.

h. _____ These fontanels are situated at the junction of the parietal, occipital, and temporal bones.

i. Refer to Figure LG 6.2 and label the visible fontanels.

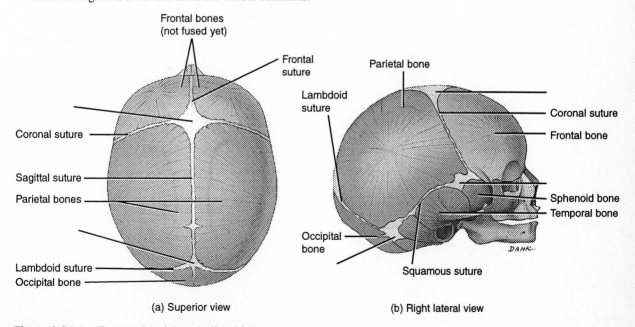

(a) Superior view   **(b) Right lateral view**

Figure LG6.2   Fontanels of the skull at birth.

D5. Match the cranial bone to the feature associated with it (you may use the answers more than once).

E ethmoid bone	P parietal bone
F frontal bone	S sphenoid bone
O occipital bone	T temporal bone

a. _____ sella turcica

b. _____ foramen magnum

c. _____ foramen ovale

d. _____ petrous portion

e. _____ supraorbital foramen

f. _____ mastoid process

g. _____ greater wings

h. _____ hypoglossal arch

i. _____ jugular foramen

j. _____ superior nasal conchae

k. _____ external auditory meatus

l. _____ optic foramen

m. _____ mandibular fossa

n. _____ zygomatic process

o. _____ cribriform plate

p. _____ superior nuchal line

q. _____ lateral masses

r. _____ carotid foramen

s. _____ metopic suture

t. _____ perpendicular plate

u. _____ lesser wing

v. _____ pterygoid processes

w. _____ olfactory foramina

x. _____ crista galli

y. _____ styloid process

D6. Check your knowledge of the facial bones by filling in the blanks.

a. The paired _____ bones meet at the middle and superior part of the face and form part of the bridge of the nose.

b. The two _____ bones commonly are called cheek bones.

c. The _____ bones are the smallest bones in the face.

d. The bones that promote turbulent filtration and circulation of air before entering the lungs

are the _____ _____ _____.

e. The _____ forms the inferior and posterior portion of the nasal septum.

f. The mental foramen is associated with the _____.

g. The paired _____ unite to form the upper jawbone.

h. The L-shaped bones that form the posterior portion of the hard palate are called the

_____ bones.

D7. Match the feature listed below with the correct facial bone.

MA	mandible
MX	maxilla bone
P	palatine bone
Z	zygomatic bone

a. _____ horizontal plates   c. _____ temporal process

b. _____ coronoid process   d. _____ palatine process

D8. Give the location for the opening of each orbit listed below.

a. Optic foramen

b. Superior orbital fissure

c. Inferior orbital fissure

d. Supraorbital foramen

e. Canal for nasolacrimal duct

D9. Give the location of and structures passing through the following foramina (see Table 6.4).

Foramen	Location	Structures
Carotid		
Jugular		
Lacerum		
Magnum		
Mandibular		
Mental		
Olfactory		
Optic		
Ovale		
Rotundum		
Spinosum		

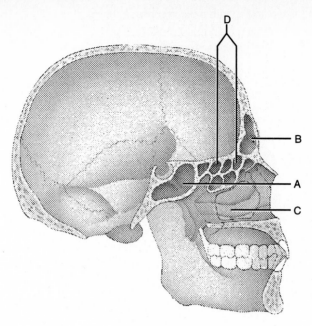

Median view

Figure LG 6.3   Paranasal sinuses.

D10. Refer to Figure LG 6.3 and label the following structures.

_____ ethmoidal sinus   _____ maxillary sinus

_____ frontal sinus   _____ sphenoidal sinus

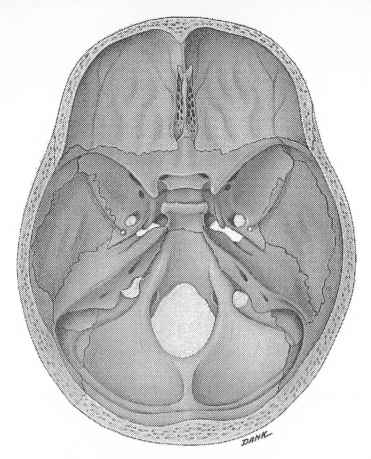

Figure LG 6.4 Sphenoid bone and cranial fossae.

D11. Refer to Figure LG 6.4; with a colored pencil, shade in the following structures. Be sure that your shading corresponds with the color code ovals.

○ ethmoid bone   ○ sphenoid bone

○ occipital bone   ○ temporal bone

○ parietal bone

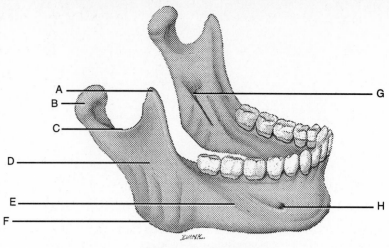

Diagram of right lateral view

Figure LG 6.5 Mandible.

D12. Refer to Figure LG 6.5 and label the following structures of the mandible.

_____ angle   _____ mandibular foramen

_____ body   _____ mandibular notch

_____ condylar process   _____ mental foramen

_____ coronoid process   _____ ramus

E. Hyoid Bone (page 149)

E1. List the basic features of the hyoid bone and give its function.

F. Vertebral Column (pages 150–158)

F1. Complete the following statements.

a. The vertebral column makes up about _____ of the total height of the body.

b. The openings between vertebrae are called _____ _____.

c. The adult vertebral column contains _____ vertebrae distributed as _____ cervical

vertebrae, _____ thoracic vertebrae, _____ lumbar vertebrae, _____ sacrum, and

_____ coccyx.

d. Fibrocartilaginous _____ _____ are located between vertebrae from C2–3 through L5–S1.

e. The adult vertebral column has _____ normal curves.

f. The _____ and _____ curves are called secondary

    curves, while the _____ and _____ curves are called

    primary curves.

g. The two portions of a disc are the _____ _____ and

    the _____ _____.

h. Prior to the fusion of the sacral and coccygeal vertebrae, the total number of vertebrae in the

    human spine is _____.

F2. Refer to Figure LG 6.6, then label and color the vertebrae in each spinal region. Be sure
    to select the same color for the corresponding color code oval.
    ○ cervical
    ○ thoracic
    ○ lumbar
    ○ sacral

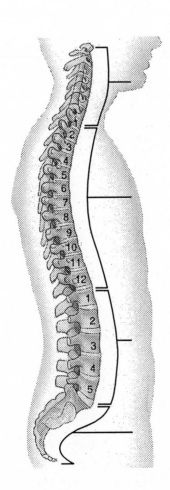

Figure LG 6.6   Vertebral column.

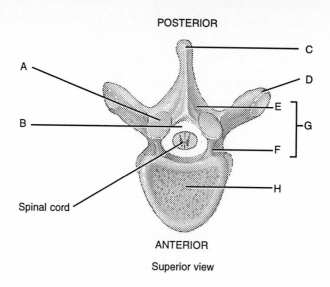

POSTERIOR

A

B

C

D

E

G

F

H

Spinal cord

ANTERIOR

Superior view

Figure LG 6.7   Typical vertebra.

F3.   Refer to Figure LG 6.7 and complete the following actions.

a.   Label the following structures.

_____ body ⭘

_____ lamina ⭘

_____ pedicle ⭘

_____ spinous process ⭘

_____ superior articular process ⭘

_____ transverse process ⭘

_____ vertebral arch ⭘

_____ vertebral foramen ⭘

b.   Color the structures listed above. Be sure that your color corresponds to the color code ovals.

F4.   List the foramina found in all cervical vertebrae and the structures that are found there.

a.

b.

F5.   Contrast the features of the atlas (C1), axis (C2), and the vertebra prominens (C7).

F6.   What is the function of the demifacets found on thoracic vertebrae?

F7. Check your knowledge of the sacrum and coccyx. Answer (T) true or (F) false to the following questions.

a. _____ Fusion begins between 13 and 15 years of age.

b. _____ The sacral canal is a continuation of the vertebral canal.

c. _____ The inferior entrance to the vertebral canal is called the sacral promontory.

d. _____ The median sacral crest occurs when the spinous processes of the sacral segments fuse together.

e. _____ The coccyx is initially composed of three segments.

f. _____ Fusion of the coccyx generally occurs between 20 and 30 years of age.

G. Thorax (pages 158–162)

G1. Refer to Figure LG 6.8 and label the following structures.

_____ body

_____ clavicular notch

_____ manubrium

_____ sternal angle

_____ suprasternal notch

_____ xiphoid process

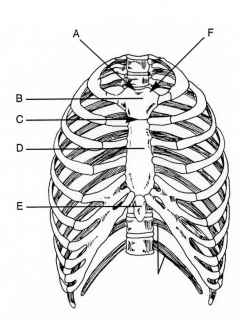

Figure LG 6.8   Anterior view of the thorax.

G2. Test your knowledge of the ribs using the answers in the box below.

body	floating ribs
costal angle	head
costal cartilage	intercostal spaces
costal groove	true ribs
false ribs	vertebrochondral ribs

a. The _____ of a typical rib is a projection at the posterior end.

b. The _____ _____ are occupied by intercostal muscles, blood vessels, and nerves.

c. The _____ is the main part of the rib, while the _____

_____ is the point where the rib takes an abrupt change in curvature.

d. The inner surface of the rib has a _____ _____ that protects blood vessels and a small nerve.

e. The thin strip of hyaline cartilage that attaches the first through seventh ribs to the sternum

is called the _____ _____.

f. The first seven pairs of ribs are called _____ _____

and the remaining five pairs are called _____ _____.

g. The eleventh and twelfth ribs are designated _____ _____.

h. The eighth, ninth, and tenth ribs are classified as _____.

H. Applications to Health (Refer to *Applications to Health* companion to answer Part H questions.)

H1. Describe the phenomenon of a herniated (slipped) disc..

H2. Contrast scoliosis, kyphosis, and lordosis

H3. Spina bifida involves a nonunion defect of the _____, leading to a protrusion of the membranes around the spinal cord or the cord itself.

H4. The most frequent fractures of the vertebrae involve _____ and

_____.

ANSWERS TO SELECT QUESTIONS

A1. 80, 126.

A2. girdles.

B1. (a) 6; (b) 3; (c) 1; (d) 5; (e) 2; (f) 4; (g) 3; (h) 1; (i) 4.

C1. Ligament and muscle attachment; grooves for blood vessels; openings for nerves and blood vessels.

C2.

Marking	Description	Example
Fissure	Narrow slit between parts of bones through which blood vessels or nerves pass	Superior orbital fissure of the sphenoid
Foramen	Opening through which blood vessels, nerves, and ligaments pass	Infraorbital foramen of the maxilla
Sulcus	A furrow that accommodates a soft structure such as a blood vessel, nerve, or tendon	Intertuberculus sulcus of the humerus
Fossa	A depression in or on a bone	Mandibular fossa of the temporal bone
Facet	A smooth, flat surface	Articular facet for the tubercle of rib on a vertebra
Tuberosity	A large, rounded, usually roughed process	Greater tubercle of the humerus
Trochanter	A large projection found only on the femur	Greater trochanter of the femur
Crest	A prominent ridge on a bone	Iliac crest of the hip-bone
Spinous process	A sharp, slender process	Spinous process of a vertebrae
Epicondyle	A prominence above a condyle	Medial epicondyle of the femur

D3. (a) 22; (b) cranial, protect; (c) 14; (d) suture; (e) fontanels.

D4. (a) 2; (b) 4; (c) 5; (d) 3; (e) 7; (f) 1; (g) 8; (h) 6.

D5. (a) S; (b) O; (c) S; (d) T; (e) F; (f) T; (g) S; (h) O; (i) T; (j) E; (k) T; (l) S; (m) T; (n) T; (o) E; (p) O; (q) E; (r) T; s) F; (t) E; (u) S; (v) S; (w) E; (x) E; (y) T.

D6. (a) Nasal; (b) zygomatic; (c) lacrimal; (d) inferior nasal conchae; (e) vomer; (f) mandible; (g) maxillae; (h) palatine.

D7. (a) P; (b) MA; (c) Z; (d) MX.

D9.

Foramen	Location	Structures
Carotid	Temporal	Internal carotid artery
Jugular	Temporal	Internal jugular vein, cranial nerves IX, X, XI
Lacerum	Sphenoid	Branch of ascending pharyngeal artery
Magnum	Occipital	Medulla oblongata, CN XI, vertebral and spinal arteries
Mandibular	Mandible	Inferior alveolar nerves and vessels
Mental	Mandible	Mental nerves and vessels
Olfactory	Ethmoid	CN I (olfactory)
Optic	Sphenoid	CN II (optic)
Ovale	Sphenoid	CN V (Mandibular branch)
Rotundum	Sphenoid	CN V (Maxillary branch)
Spinosum	Sphenoid	Midle meningeal blood vessels

D10. A—sphenoidal sinus, B—frontal sinus, C—maxillary sinus, D—ethmoidal sinus.

D12. A—coronoid process, B—condylar process, C—mandibular notch, D—ramus, E—body, F—angle, G—Mandibular foramen, H—mental foramen.

F1. (a) Two-fifths; (b) intervertebral foramina; (c) 26, 7, 12, 5, 1, 1; (d) intervertebral discs; (e) 4; (f) cervical, lumbar; thoracic, sacral; (g) annulus fibrosus, nucleus pulposus; (h) 33.

F3. A—Superior articular process, B—vertebral foramen, C—spinous process, D—transverse process, E—lamina, F—pedicle, G—vertebral arch, H—body.

F4. (a) Vertebral foramen, spinal cord; (b) transverse foramen, vertebral artery.

F6. Articulate with the heads of the ribs.

F7. (a) F; (b) T; (c) F; (d) T; (e) F; (f) T.

G1. A—suprasternal notch, B—manubrium, C—sternal angle, D—body, E—xiphoid process, F—Clavicular notch.

G2. (a) Head, (b) intercostal spaces; (c) body, costal angle; (d) costal groove; (e) costal cartilage; (f) true ribs, false ribs; (g) floating ribs; (h) vertebrochondral.

H3. Laminae.

H4. T5–T6, T9–T12.

SELF QUIZ

Choose the one best answer to the following questions.

1. All of these bones contain paranasal sinuses except

   A. frontal
   B. maxilla
   C. nasal
   D. sphenoid
   E. ethmoid

2. A baby born with a cleft palate results from the failure of which of the following bones to fuse?

   A. maxillae
   B. mandible
   C. vomer
   D. inferior nasal conchae
   E. zygomatic

3. Which of the following structures is NOT a part of the ethmoid bone?

   A. crista galli
   B. perpendicular plate
   C. olfactory foramen
   D. cribriform plate
   E. horizontal plate

4. Which of these pairs is NOT matched correctly?

   A. sternum—xiphoid
   B. axis—dens
   C. sphenoid—sella turcica
   D. temporal bone—mastoid
   E. zygomatic—alveoli

5. The site of attachment of the meninges is the

   A. sphenoid bone
   B. crista galli of ethmoid
   C. sella turcica
   D. vomer
   E. inferior nasal conchae

6. A lateral curvature of the spine is called

   A. scoliosis
   B. kyphosis
   C. lordosis
   D. "humpback"
   E. spina bifida

7. Which suture is located between the temporal and parietal bones?

   A. lambdoid
   B. coronal
   C. squamous
   D. sagittal
   E. none of the above are correct

8. The "soft spot" in a newborn's skull is called the

   A. foramen
   B. fossa
   C. meatus
   D. fontanel
   E. sulcus

9. All the following are bones of the axial skeleton except

   A. rib
   B. sternum
   C. coccyx
   D. hyoid
   E. clavicle

10. Which bone forms the inferior portion of the nasal septum?

    A. ethmoid
    B. inferior nasal concha
    C. vomer
    D. palatine
    E. none of the above are correct

11. The mental foramen is associated with this bone.

    A. maxilla
    B. mandible
    C. temporal
    D. palatine
    E. sphenoid

12. The most inferior portion of the sternum is called the _____.

    A. manubrium
    B. sternal angle
    C. body
    D. xiphoid process
    E. none of the above are correct

13. Which part of the sacrum represents the fusion of the transverse processes?

    A. auricular surface
    B. ala
    C. median sacral crest
    D. lateral sacral crest
    E. transverse line

14. The hard palate is composed of _____ bones.

    A. two maxillae
    B. two maxillae and two mandible
    C. two maxillae and two palatine
    D. two palatine
    E. ethmoid, vomer, and palatine

15. A foramen is

    A. a cavity within a bone
    B. a depression
    C. a hole for blood vessels and nerves
    D. a ridge
    E. a bony prominence

Answer (T) True or (F) False to the following questions.

16. _____ The cervical and lumbar curves of the spine are primary curves.

17. _____ The inner surface of a rib has a costal space that protects blood vessels and a small nerve.

18. _____ A kyphosis is an exaggeration of the lumbar curve.

19. _____ The superior and inferior surfaces of the vertebral body attach to the intervertebral discs.

20. _____ The mandibular branch of the trigeminal (V) nerve passes through the foramen ovale.

Fill in the blanks.

21. A fingerlike or toothlike projection called the _____ is part of the axis.

22. The _____ bone is frequently fractured during strangulation.

23. The _____ fontanel is the last one to close.

24. The _____ suture is located between the parietal and occipital bones.

25. There are _____ bones in the axial skeleton.

ANSWERS TO THE SELF QUIZ

1. C	10. C	19. T
2. A	11. B	20. T
3. E	12. D	21. Dens or odontoid process
4. E	13. D	22. Hyoid
5. B	14. C	23. Anterior
6. A	15. C	24. Lambdoid
7. C	16. F	25. 80
8. D	17. F	
9. E	18. F	

The Skeletal System: The Appendicular Skeleton

SYNOPSIS

This chapter continues our discussion of the skeletal system. The 126 bones of the **appendicular skeleton** comprise the **pectoral** and **pelvic girdles** and the upper and **lower limbs** (extremities). The wide array of movements possible with our bodies is produced by the highly coordinated actions of the appendicular and axial skeletons, their joints, and the muscular system.

TOPIC OUTLINE AND OBJECTIVES

A. Pectoral (Shoulder) Girdle

1. Identify the bones of the pectoral (shoulder) girdle and major features.

B. Upper Limb (Extremity)

2. Identify the bones of the arm, forearm, wrist, hand, and fingers, and their attributes.

C. Pelvic (Hip) Girdle

3. Identify the components of the pelvic (hip) girdle and their principal markings.

D. Female and Male Skeletons

4. Contrast the structural variation between the female and male skeletons.

E. Lower Limb (Extremity)

5. Identify the bones of the thigh, knee, leg, ankle, foot, and toes and their markings.
6. Discuss the characteristics of the arches of the foot.

SCIENTIFIC TERMINOLOGY

Find an anatomical sample word for each prefix and suffix:

Prefix/Suffix	Meaning	Sample Word
acro-	top or summit	
arthro-	joint	
auricula-	ear shaped	
clavis-	key	
equinus-	horse	
hemi-	one half	
inter-	between	
korne-	crown shaped	
konos-	cone	
meta-	after or beyond	
pes-	foot	
-plasty	surgical repair	

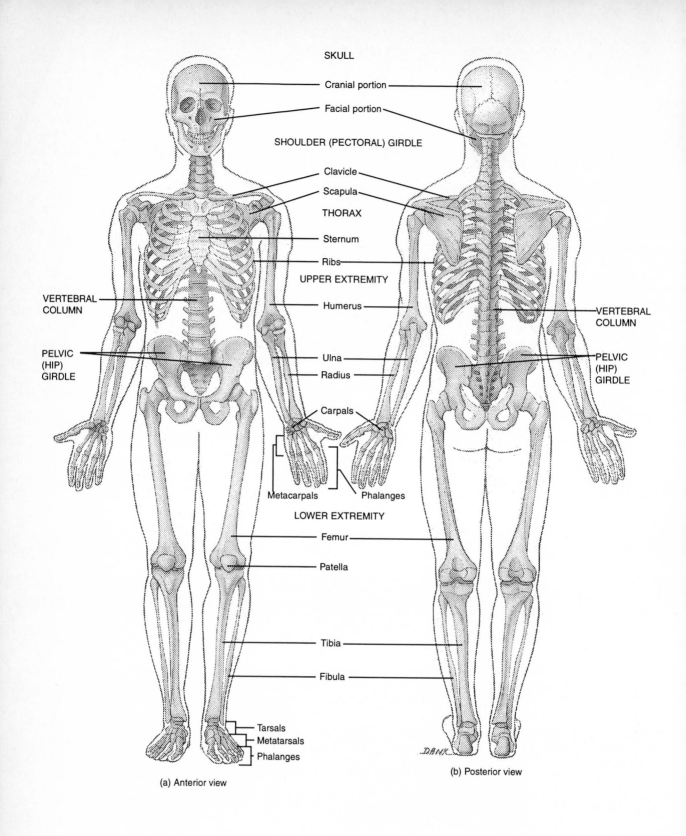

SKULL
Cranial portion
Facial portion

SHOULDER (PECTORAL) GIRDLE
Clavicle
Scapula
THORAX
Sternum
Ribs
UPPER EXTREMITY
Humerus

VERTEBRAL COLUMN

VERTEBRAL COLUMN

Ulna
Radius

PELVIC (HIP) GIRDLE

PELVIC (HIP) GIRDLE

Carpals

Metacarpals   Phalanges

LOWER EXTREMITY
Femur
Patella

Tibia

Fibula

Tarsals
Metatarsals
Phalanges

(a) Anterior view

(b) Posterior view

Figure LG 7.1   Divisions of the skeletal system.

A. Pectoral (Shoulder) Girdle   (pages 167–168)

A1. Fill in the blanks for the questions relating to the pectoral girdle. Refer to Figure LG 7.1 to check your answers.

a. The pectoral girdle consists of two bones: the _____ and

_____.

b. The medial end of the clavicle articulates with the _____.

c. The broad, flat, lateral end of the clavicle, the _____

_____, articulates with the scapula. The joint is called the

_____ joint.

d. The _____ _____ on the inferior surface of the lateral
end serves as a point of attachment for a ligament.

e. The clavicle is one of the *(most? least?)* frequently fractured bone in the body.

A2. Check your knowledge of the scapula; choose the correct answer.

a. The scapulae are large *(rectangular? triangular?)* flat bones situated in the posterior part of
the thorax between the levels of the *(2nd–7th? 3th–8th?)* ribs.

b. A sharp ridge, the *(body? spine?),* runs diagonally across the posterior surface.

c. The *(acromion? glenoid cavity?)* articulates with the head of the humerus.

d. At the lateral end of the superior border is a projection of the anterior surface called the
(conoid? coracoid?) process.

e. On the ventral surface of the scapula is a lightly hollowed-out area called the *(supraspinatus? subscapular?)* fossa.

B. Upper Limb (Extremity)   (pages 168–175)

B1. Test your knowledge of the upper limb (extremity) by answering the questions below.
Refer to your textbook and Figure LG 7.1 to check your answers.

a. The upper limb (extremities) consist of _____ bones.

b. The _____ is the longest and largest bone in the upper extremity. Its

distal end articulates with the _____ and the _____.

c. Between the greater and lesser tubercles runs a furrow called the

_____ _____.

d. The roughened V-shaped area called the _____ _____
serves as a point of attachment for the deltoid muscle.

B2. At the distal end of the humerus are several structures. Match the structure with the correct description.

CA capitulum	RF radial fossa
CO coronoid fossa	T trochlea
O olecranon fossa	

a. _____ Posterior depression.

b. _____ Rounded knob that articulates with the head of the radius.

c. _____ Receives ulna when forearm is flexed.

d. _____ Spool-shaped surface that articulates with the ulna.

e. _____ Receives head of radius when forearm is flexed.

B3. Test your knowledge of the radius and ulna.

a. The _____ is the medial forearm bone, while the

_____ is located on the lateral side.

b. At the proximal end of the ulna is the _____, which forms the prominence of the elbow.

c. The _____ _____ is an anterior projection of the ulna that, together with the olecranon, receives the trochlea of the humerus.

d. The _____ _____ is a curved area between the olecranon and the coronoid process.

e. A _____ _____ is on the posterior side of the distal end of the ulna.

f. At the distal end of the radius are two structures. On the lateral side is the

_____ _____ and on the medial side is the

_____ _____.

g. The disc-shaped structure that articulates with the capitulum of the humerus is the

_____.

h. The biceps brachii muscle attaches to the _____

_____.

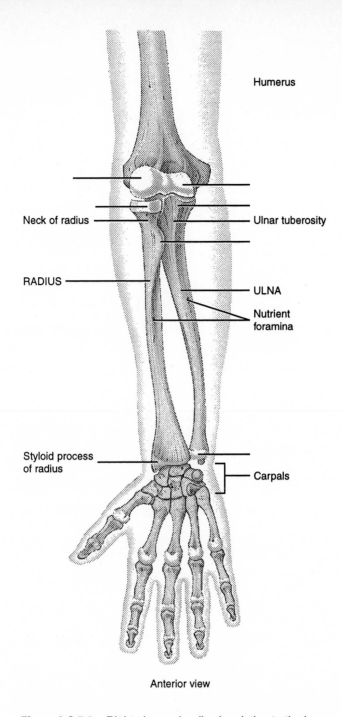

Humerus

Neck of radius

Ulnar tuberosity

RADIUS

ULNA

Nutrient
foramina

Styloid process
of radius

Carpals

Anterior view

Figure LG 7.2   Right ulna and radius in relation to the humerus and carpals.

B4.   Refer to Figure LG 7.2 and answer these questions.

a.   Label the capitulum, head of radius, trochlea, coronoid process, radial tuberosity, and the head of ulna.

b.   Color the humerus, radius, ulna, and carpals different colors.

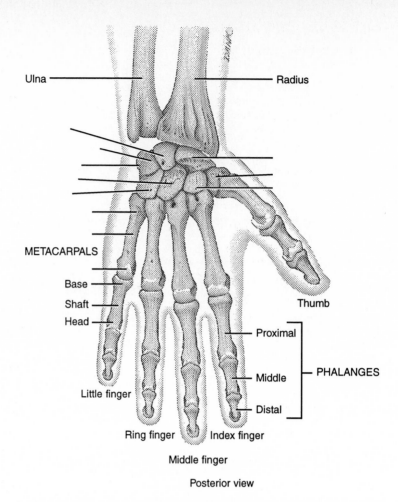

Ulna — — Radius

METACARPALS

Base —

Shaft —

Head —

Thumb

Proximal

Little finger

Middle

PHALANGES

Distal

Ring finger   Index finger

Middle finger

Posterior view

Figure LG 7.3   Right wrist and hand in relation to the ulna and radius.

B5.   Refer to Figure LG 7.3 and do the following actions.

a. Label the carpal bones.

b. Label the parts of the metacarpals.

c. Color the proximal, middle, and distal phalanges different colors.

C. Pelvic (Hip) Girdle   (pages 175–178)

C1. Complete these statements for the pelvic (hip) girdle.

a. The pelvic girdle consists of two hipbones or _____ bones.

b. The hipbones unite to each other anteriorly at a joint called the _____

_____.

c. The hipbones of a newborn consist of three components: a superior

_____, inferior and anterior _____, and an inferior and

posterior _____.

d. The deep fossa that articulates with the head of the femur is the _____.

e. The _____ _____ articulates with the sacrum to form
the sacroiliac joint.

f. The superior border of the ilium is the _____ _____. It

ends anteriorly in the _____ _____

_____ spine and ends posteriorly in the _____

_____ _____ -spine.

g. The portion of the bony pelvis above the pelvic brim is called the _____

_____.

h. The pelvic inlet in females is *(larger and more oval? heart shaped?)* compared to that in
males.

i. Below the posterior inferior iliac spine is the_____

_____ _____ , through which the longest nerve in the
body passes.

D. Female and Male Skeletons   (page 178)

D1. In the table below, contrast the differences between the female and male pelvis.

Point of Comparison	Female	Male
General structure		
Greater sciatic notch		
Greater pelvis		
Pelvic inlet		
Pelvic outlet		
Iliac crest		
Ilium		
Pubic arch		

E. Lower Limb (Extremity)   (pages 178–186)

E1.   Complete the questions below pertaining to the femur.

a. The lower extremities are composed of _____ bones.

b. The proximal rounded _____ of the femur articulates with the acetabulum of the hipbone.

c. The region where fairly common fractures in the elderly occur is the

   _____ of the femur.

d. The _____ _____ and _____

   _____ are projections that serve as points of attachment for some of the thigh and buttock muscles.

e. The vertical ridge on the posterior surface of the shaft is the _____

   _____ .

f. A depressed area between the condyles on the posterior surface is called the

   _____ _____ .

g. A (decreased? increased?) angle of the neck of the femur produces a "knock-knee" condition.

E2.   Complete the exercise for the patella, tibia, and fibula. Refer to your textbook and Figure LG 7.1 to check your answers.

a. The distal projection of the fibula is called the _____

   _____ .

b. The _____ _____ on the anterior surface of the tibia is a point of attachment for the patellar ligament.

c. The _____ , or shinbone, is on the _____ side of the leg.

d. The broad superior end of the patella is called the _____ .

e. The medial surface of the distal end of the tibia forms the _____

   _____ .

E3.   Complete the exercise for the tarsals, metatarsals, and phalanges.

a. The _____ is the uppermost tarsal bone and is the only bone of the foot that articulates with the fibula and tibia.

b. The _____ is the largest and strongest tarsal bone.

c. The head of the metatarsus articulates with the _____ row of phalanges.

d. There are _____ tarsal bones, _____ metatarsals, and _____ phalanges found in each foot.

e. The great or big toe is also called the _____ .

E4. List the three functions of the arches of the foot.

a.

b.

c.

E5. Explain the following terms.

a. Flatfoot

b. Clawfoot

c. Clubfoot

d. Bunion

F. Summary

F1. Match the common name with the scientific name.

a. _____ shoulder blade   1. humerus

b. _____ collar bone   2. phalanges

c. _____ elbow   3. tarsals

d. _____ wrist   4. radial & ulna

e. _____ kneecap   5. scapula

f. _____ thigh   6. carpals

g. _____ shin   7. patella

h. _____ fingers/toes   8. femur

i. _____ arm   9. tibia

j. _____ forearm   10. olecranon

k. _____ ankle   11. clavicle

ANSWERS TO SELECT QUESTIONS

A1. (a) Clavicle, scapula; (b) sternum; (c) acromial extremity, acromioclavicular; (d) conoid tubercle; (e) most.

A2. (a) Triangular, 2nd–7th; (b) spine; (c) glenoid cavity; (d) coracoid; (e) subscapular.

B1. (a) 60; (b) humerus, radius, ulna; (c) intertubercular sulcus; (d) deltoid tuberosity.

B2. (a) O; (b) CA ; (c) CO (d) T; (e) RF.

B3. (a) Ulna, radius; (b) olecranon; (c) coronoid process; (d) trochlear notch; (e) styloid process; (f) styloid process; ulnar notch (g) head; (h) radial tuberosity.

C1. (a) Coxal; (b) pubic symphysis; (c) ilium, pubis, ischium; (d) acetabulum; (e) auricular surface; (f) iliac crest, anterior superior iliac, posterior superior iliac; (g) greater pelvis; (h) larger and more oval; (i) greater sciatic notch.

D1.

Point of Comparison	Female	Male
General structure	Light and thin	Heavy and thick
Greater sciatic notch	Wide	Narrow
Greater pelvis	Shallow	Deep
Pelvic inlet	Larger and oval	Heart shaped
Pelvic outlet	Wider	Narrower
Iliac crest	Less curved	More curved
Ilium	Less vertical	More vertical
Pubic arch	Greater than 90°	Less than 90°

E1. (a) 60; (b) head; (c) neck; (d) greater trochanter, lesser trochanter; (e) linea aspera; (f) intercondylar fossa; (g) decreased.

E2. (a) Lateral malleolus; (b) tibial tuberosity; (c) tibia, medial; (d) base; (e) medial malleolus.

E3. (a) Talus; (b) calcaneus; (c) proximal; (d) 7, 5, 14; (e) hallux.

E4. (a) support body weight; (b) distribute body weight; (c) provide leverage for walking.

F1. (a) 5; (b) 11; (c) 10; (d) 6; (e) 7; (f) 8; (g) 9; (h) 2; (i) 1; (j) 4; (k) 3.

SELF QUIZ

Choose the one best answer to the following questions.

1. The femur's articulation to the pelvic bone is the

   A. glenoid fossa
   B. acetabulum
   C. linea aspera
   D. olecranon
   E. trochlea

2. Which of the following bones is NOT part of the appendicular skeleton?

   A. humerus
   B. patella
   C. talus
   D. pisiform
   E. atlas

3. The humerus articulates with all of the following bones except the

   A. scapula
   B. radius
   C. ulna
   D. clavicle
   E. none of the above are correct

4. Which of the following is mismatched?

   A. scapula— acromion
   B. ulna—olecranon
   C. tibia—malleolus
   D. fibula—styloid process
   E. pelvis—pubis

5. Which of the following statements concerning the pelvis is incorrect?

   A. The iliac crest is less curved in females than in males.
   B. The male pelvic inlet is heart shaped.
   C. The pubic arch in females is less than a 90° angle.
   D. The greater pelvis is deeper in the male than in the female.
   E. All of the above statements are correct.

6. Distally, the tibia articulates with the

   _____.

   A. patella
   B. femur
   C. calcaneus
   D. talus
   E. navicular

7. The coracoid process is associated with which bone?

   A. ulna
   B. scapula
   C. femur
   D. fibula
   E. humerus

8. The lateral end of the clavicle articulates with the

    A. sternum
    B. spine of scapula
    C. acromion of scapula
    D. lateral border of scapula
    E. glenoid of scapula

9. The portion of the metacarpals that form the "knuckles" is the

    A. base
    B. shaft
    C. head
    D. phalanx
    E. none of the above are correct

10. The greater trochanter is associated with this bone.

    A. humerus
    B. fibula
    C. tibia
    D. femur
    E. ulna

Answer (T) True or (F) False to the following questions.

11. _____ The scapula articulates with the ribs.

12. _____ The ulna is the lateral bone of the forearm.

13. _____ The patella articulates with the tibia.

14. _____ The distal fibula articulates only with the talus.

15. _____ The calcaneus is more commonly known as the heel bone.

16. _____ There are 14 phalanges in each hand and also in each foot.

17. _____ The trochlear notch is associated with the ulna.

18. _____ The medial end of the clavicle forms a joint called the sternoclavicular joint.

19. _____ The upper extremities consist of 62 bones.

20. _____ The coronoid process is located on the scapula.

Fill in the blanks.

21. The _____ is one of the most frequently fractured bones in the body.

22. The _____ is located in the fusion area of the ischium, pubis, and ilium.

23. The _____ of the patella is the pointed, inferior end.

24. The _____ is the thinnest bone in the body compared to its length.

25. The carpal bone that derives its name from its pea shape is the _____.

ANSWERS TO THE SELF QUIZ

1. B	10. D	19. F
2. E	11. F	20. F
3. D	12. F	21. Clavicle
4. D	13. F	22. Acetabulum
5. C	14. F	23. Apex
6. C	15. T	24. Fibula
7. B	16. T	25. Pisiform
8. C	17. T	
9. C	18. T	

Joints

SYNOPSIS

The complex organization of bone tissue and the variety of bones associated with the skeleton are not enough to make it a truly functional system. The **joints** provide the key ingredient necessary to complete this system.

A **joint** or **articulation** occurs wherever two bones come together. Once again you will see that anatomy greatly influences physiology, as the function of each joint is dependent upon its anatomy. Some joints permit a wide triaxial range of motion while others have a limited monaxial motion. It is upon this relationship that joints can be classified, by either their function or structure.

Joints, however, are more than the meeting of two bones. An array of accessory structures including **ligaments**, **cartilage**, **synovial membranes** and **fluid**, and **bursae** are often associated with joints.

Despite the excellent design of joints, joint injuries still persist as one of the most common skeletal maladies.

TOPIC OUTLINE AND OBJECTIVES

A. Classification of Joints

1. Define a joint and identify the factors that determine the degree of movement at a joint.
2. Differentiate between the structural and functional classifications of joints.

B. Fibrous Joints

3. Contrast the three fibrous joints: sutures, syndesmoses, and gomphoses.

C. Cartilaginous Joints

4. Contrast the two cartilaginous joints: synchondroses and symphyses.

D. Synovial Joints

5. Define a synovial joint and identify its characteristics.
6. Discuss the accessory structures associated with joints.
7. Describe and compare the types of movement possible at synovial joints.

E. Selected Joints of the Body

F. Aging and Joints

G. Applications to Health

8. Define and describe rheumatism, arthritis, Lyme disease, bursitis, ankylosing spondylitis, dislocation, sprain, and strain.

H. Key Medical Terms Associated with Joints

SCIENTIFIC TERMINOLOGY

Find an anatomical sample word for each prefix and suffix:

Prefix/Suffix	Meaning	Sample Word
-algia	pain	
amphi-	on both sides	
arthro-	joint	
chondro-	cartilage	
circum-	around	
-ectomy	removal of	
hyper-	excessive	
-kinesis	movement	
rheuma-	discharge	
spondyl-	vertebra	
syn-	together	
syndesmo-	band or ligament	

A. Classification of Joints   (page 190)

A1.   Complete the following questions about joint classification.

a. _____ joints have _____ joint cavity and the bones are

held together by fibrous (collagenous) _____ tissue.

b. _____ joints possess a joint cavity and the bones forming the joints are

united by a surrounding _____ capsule.

c. A _____ is an immovable joint.

d. A freely movable joint is a _____.

e. _____ joints have no joint cavity, and the bones are held together by
cartilage.

f. A(n) _____ is a partially movable joint.

B. Fibrous Joints   (pages 190–191)

B1.   Classify the joints listed below (you may use the answers more than once).

G gomphosis	SUT sutures	SYN syndesmoses

a. _____ Joint in which a cone-shaped peg fits into a socket.

b. _____ Found between the bones of the skull.

c. _____ An example is the distal articulation of the tibia and fibula.

d. _____ The articulations of the roots of the teeth with the alveoli of the maxillae and
mandible.

e. _____ Has considerably more fibrous connective tissue than a suture.

f. _____ When completely fused, they are called synostoses.

C. Cartilaginous Joints (page 191)

C1. Define and give an example of the two types of cartilaginous joints.

a.

b.

D. Synovial Joints (pages 191–201)

D1. On Figure LG 8.1, label and color the indicated structures. Make sure that your colors correspond with the color code ovals.

- ◯ articular bone
- ◯ articular cartilage
- ◯ fibrous capsule

- ◯ periosteum
- ◯ synovial (joint) cavity
- ◯ synovial membrane

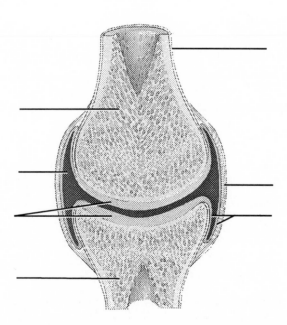

Figure LG 8.1   Synovial joint. Generalized structure.

D2. Test your knowledge of the synovial joints by answering the questions below.

a. A synovial joint is functionally classified as a _____.

b. Synovial joints are characterized by a _____ cavity and the presence of

_____ cartilage.

c. Synovial joints are surrounded by a sleevelike articular _____, which is

composed of two layers: an outer _____ _____ and an

inner _____ _____.

d. It is the strength of the *(ligaments? tendons?)* that hold bone to bone.

e. The liquid called _____ _____ lubricates the joint and provides nourishment for the articular cartilage.

D3. Synovial joints contain accessory ligaments and structures. Complete the following questions about these structures.

a. Contrast extracapsular and intracapsular ligaments.

b. What is an articular disc (meniscus)? Give an example of where this structure can be found.

c. Describe the structure and function of a bursa.

D4. List six factors that affect movement of a synovial joint.

a.

b.

c.

d.

e.

f.

D5. Define the following terms. Give an example for each type of movement.

a. Gliding movement

b. Flexion

c. Extension

d. Abduction

e. Adduction

f. Rotation

g. Circumduction

D6. Match the special movement with the correct definition.

a. _____ inversion

b. _____ eversion

c. _____ dorsiflexion

d. _____ plantar flexion

e. _____ protraction

f. _____ retraction

g. _____ supination

h. _____ pronation

i. _____ elevation

j. _____ depression

1. a downward (inferior) movement of a part of the body

2. movement of the forearm in which the palm is turned posteriorly or inferiorly

3. an example is pulling the lower jaw back in line with the upper jaw (a movement back to anatomical position)

4. bending the foot at the ankle joint in the direction of the plantar surface

5. movement of the sole of the foot medially

6. an upward (superior) movement of a part of the body

7. movement of the forearm in which the palm is turned anteriorly or superiorly

8. an example is moving the shoulder girdle anteriorly

9. bending the foot at the ankle joint in the direction of the dorsum

10. movement of the sole of the foot laterally

D7. Fill in the chart below.

Type of Joint	Description	Movement
Gliding		
	Convex surface fits into a concave surface	Monaxial
	Rounded surface fits into a ring formed by bone and ligament	
Condyloid		Biaxial
Saddle		
	Ball-like surface fits into cuplike depression	

E. Selected Joints of the Body (pages 201–219)

E1. Refer to Figure LG 8.2 and color the following structures. Be sure that your colors correspond with the color code ovals.

○ acromioclavicular ligament

○ articular capsule

○ coracohumeral ligament

○ glenohumeral ligament

○ subscapular bursa

○ transverse ligament

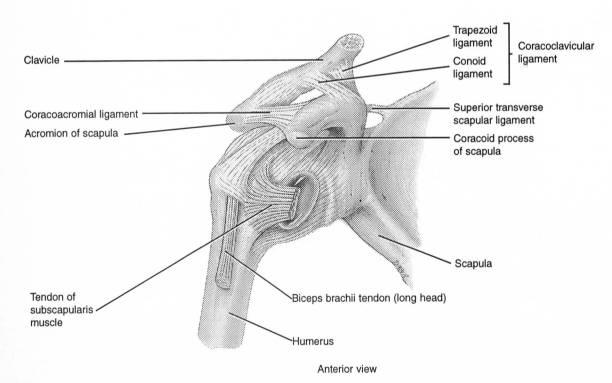

Anterior view

Figure LG 8.2 Shoulder (humeroscapular) joint.

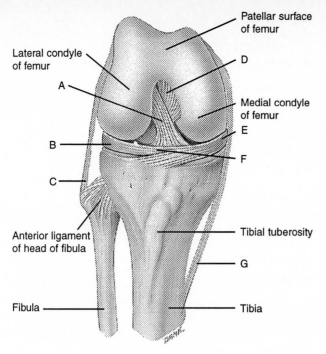

Lateral condyle of femur

Patellar surface of femur

D

A

Medial condyle of femur

E

B

F

C

Anterior ligament of head of fibula

Tibial tuberosity

G

Fibula

Tibia

Anterior deep view (flexed)

Figure LG 8.3 Knee joint.

E2. Label the following structures on Figure LG 8.3 (anterior view of the flexed knee).

_____ anterior cruciate ligament   _____ posterior cruciate ligament

_____ fibular collateral ligament   _____ tibial collateral ligament

_____ lateral meniscus   _____ transverse ligament

_____ medial meniscus

E3. Complete the following questions.

a. The _____ _____ ligament affords strength for the posterior surface of the knee joint.

b. The _____ _____ ligament is stretched or torn in 70% of all serious knee injuries.

c. Tennis elbow refers to pain at or near the lateral _____ of the humerus.

d. The _____ ligament extends from the coracoid process of the scapula to the greater tubercle of the humerus.

e. A number of ligaments provide strength and support to the knee joint. Identify whether the following ligaments are extracapsular (E) or intracapsular (I).

   1. Arcuate popliteal _____

   2. Anterior and posterior cruciate ligaments _____

   3. Tibial and fibular collateral ligaments _____

f. There are *(no? three? five?)* "more important" bursae associated with the knee joint.

g. Name three structures that begin with the letter C and are associated with the knee (the "unhappy triad").

   1. _____   2. _____   3. _____

F. Aging and Joints (page 220)

F1. List four effects of the aging process on the joints.

a.

b.

c.

d.

G. Applications to Health (Refer to *Application to Health* companion to answer Part G questions.)

G1. Differentiate between rheumatism and arthritis.

G2. Differentiate between rheumatoid arthritis, osteoarthritis, and gouty arthritis.

G3. Answer the following questions relating to Lyme disease, bursitis, dislocations, sprains, and strains.

a. Lyme disease is caused by a bacterium, _____ _____,

transmitted to humans by _____ _____. Within a few

weeks of the bite, a typical rash resembling a "_____

_____" develops.

b. A _____ is the forcible wrenching or twisting of a joint with stretching or

tearing of a ligament without dislocation, while a _____ is a stretched or
partially torn muscle.

c. A partial or incomplete dislocation is called a _____.

d. A dislocation or _____ is the displacement of a bone from a

_____, with tearing of ligaments, tendons, and _____
capsules.

e. Bursitis may be caused by _____, by an _____ or

_____ infection, or by _____

_____.

f. _____ are often associated with friction bursitis over the head of the first

metatarsal.

G4. Define ankylosing spondylitis and describe the etiology (cause), incidence, and symptoms associated with it.

ANSWERS TO SELECT QUESTIONS

A1. (a) Fibrous, no, connective; (b) synovial, articular; (c) synarthrosis; (d) diarthrosis; (e) cartilaginous; (f) amphiarthrosis.

B1. (a) G; (b) SUT; (c) SYN; (d) G; (e) SYN; (f) SUT.

C1. (a) Synchondroses; (b) symphyses.

D2. (a) Diarthrosis; (b) synovial, articular; (c) capsule, fibrous capsule, synovial membrane; (d) ligaments; (e) synovial fluid.

D3. (b) Fibrocartilage pads found between the articular surfaces in the knee joints; (c) saclike structure filled with a fluid similar to synovial fluid. Reduces friction between movable parts.

D4. (a) Structure or shape of the articulating bones; (b) strength and tension of the joint ligaments; (c) arrangement and tension of the muscles; (d) apposition of soft parts; (e) hormones; (f) disuse.

D6. (a) 5; (b) 10; (c) 9; (d) 4; (e) 8; (f) 3; (g) 7; (h) 2; (i) 6; (j) 1.

D7.

Type of Joint	Description	Movement
Gliding	**Articulating surface usually flat**	**Nonaxial**
Hinge	Convex surface fits into a concave surface	Monaxial
Pivot	Rounded surface fits into a ring formed by bone and ligament	**Monaxial**
Condyloid	**Oval-shaped condyle fits into an elliptical cavity**	Biaxial
Saddle	**Articular surface of one bone is saddle shaped; other is shaped like the legs of a rider**	**Biaxial**
Ball and Socket	Ball-like surface fits into cuplike depression	**Triaxial**

E2. A—anterior cruciate ligament; B—lateral meniscus; C—fibular collateral ligament; D—posterior cruciate ligament; E—medial meniscus; F—Transverse ligament; G—tibial collateral ligament.

E3. (a) Oblique popliteal; (b) anterior cruciate; (c) epicondyle; (d) coraco-humeral; (e) (1) E, (2) I, (3) E; (f) three; (g) (1) cruciate, (2) collateral; (3) cartilage.

F1. (a) decreased production of synovial fluid; (b) thinning of the articular cartilage; (c) shortening of ligaments; (d) loss of flexibility.

G3. (a) *Borrelia burgdorferia,* deer ticks, "bull's eye"; (b) sprain, strain; (c) subluxation; (d) luxation, joint, articular; (e) trauma, acute, chronic, rheumatoid arthritis; (f) bunions.

SELF QUIZ

Choose the one best answer to the following questions.

1. Which of the following is/are true regarding synovial fluid?

   A. It is secreted by the synovial membrane.
   B. It functions to lubricate and nourish the articular cartilage.
   C. It has the consistency of uncooked egg whites.
   D. All of the above statements are correct.
   E. None of the above statements are correct.

2. The intervertebral joints in the spinal column are known as

   A. sutures
   B. diarthroses
   C. symphyses
   D. synchondroses
   E. gomphoses

3. Slightly movable joints are referred to as

   A. amphiarthritic
   B. diarthritic
   C. synovial
   D. gliding
   E. none of the above are correct

4. The special movement that turns the sole of the foot medially is called

   A. pronation
   B. eversion
   C. supination
   D. inversion
   E. rotation

5. Lifting your arm laterally away from your body is

   A. adduction
   B. circumduction
   C. flexion
   D. extension
   E. abduction

6. A gomphosis is found between
   - A. parietal and occipital bones
   - B. epiphysis and diaphysis
   - C. the two pubic bones
   - D. the mandible and the teeth
   - E. the femur and tibia

7. The C-shaped cartilage pads within the knee joint are called
   - A. bursae
   - B. cruciae
   - C. menisci
   - D. capsulae
   - E. none of the above are correct

8. The elbow joint is an example of a _____ joint.
   - A. gliding
   - B. ginglymus
   - C. sellaris
   - D. condyloid
   - E. trochoid

9. A spheroid joint permits _____ movement.
   - A. monaxial
   - B. biaxial
   - C. triaxial
   - D. quadaxial
   - E. pentaxial

10. Which of the following are NOT synarthroses?
   - A. sutures
   - B. syndesmoses
   - C. gomphoses
   - D. synchondroses
   - E. none of the above are correct

Answer (T) True or (F) False to the following questions.

11. _____ Sutures are fibrous synarthrotic joints.

12. _____ Adduction is movement toward the midline of the body.

13. _____ Synovial fluid becomes less viscous when there is increased movement at a joint.

14. _____ Movement of a bone around its own axis is called rotation.

15. _____ A syndesmosis is a type of fibrous joint.

16. _____ Gliding joints are nonaxial.

17. _____ An example of a coadyloid joint is found between the radius and the carpals.

18. _____ If you stand on your toes, you are dorsiflexing your foot.

19. _____ A bursa functions to reduce friction between moving parts at a joint.

Fill in the blanks.

20. A pannus is associated with _____ _____.

21. The primary movement at a pivot joint is _____.

22. Another name for a freely movable joint is _____.

23. _____ refers to an inflammation of a synovial membrane in a joint.

24. In _____ arthritis, sodium urate crystals are deposited in the soft tissue of the joint.

25. _____ involves movement of the forearm in which the palm is turned posteriorly or anteriorly.

ANSWERS TO THE SELF QUIZ

1. D
2. C
3. A
4. D
5. E
6. D
7. C
8. B
9. C
10. B
11. T
12. T
13. T
14. T
15. T
16. T
17. T
18. F
19. T
20. Rheumatoid arthritis
21. Rotation
22. Diarthrosis
23. Synovitis
24. Gouty
25. Pronation

Muscle Tissue

CHAPTER

9

SYNOPSIS

The human skeleton standing in the corner of your anatomy laboratory is a testimony to the fact that bones and joints, though marvelous structures, do not possess the capacity to move under their own power. It is the contraction of muscle tissue that provides the power source for movement. This **skeletal** muscle, which constitutes approximately 40–50% of total body weight, is one of the most complex structures of the body. Muscle is composed of individual **myofibers** (muscle cells) and each cell is itself made up of thousands of **myofibrils**. The myofibrils are formed by thin and thick **myofilaments**, which are composed of protein molecules called **actin** and **myosin**, respectively. These thin and thick myofilaments are located in small compartments called **sarcomeres**.

Although the actual number of myofilaments within any given muscle cell varies, it has been estimated that in one muscle cell, measuring 1 cm × 1/10,000 cm, there are approximately 8,000 myofibrils. In addition, there are about 5,000 sarcomeres in one myofibril and approximately 500 thick myofilaments and 900 thin myofilaments in one sarcomere. Therefore, there are approximately 56,000,000,000 thin and thick myofilaments within one skeletal muscle cell.

Muscle tissue is involved in more than skeletal movement. The blood coursing through your arteries is pumped by a muscle. Your food is churned and pushed through the intestines by muscular contractions. Urination and defecation both occur due to similar muscular contractions. Breathing, eye movement, and labor and delivery also occur due to the contraction of muscle tissue.

In this chapter your journey continues as you travel deep into the structural composition of **skeletal**, **cardiac**, and **smooth** muscle. You will also learn how the nervous system interacts with muscle tissue and about the effects of exercise, steroids, and aging on muscle tissue.

TOPIC OUTLINE AND OBJECTIVES

A. Types of Muscle Tissue

1. Contrast the location, histology, nervous system control, and functions of the three types of muscle tissue.

B. Characteristics and Functions of Muscle Tissue

2. List and describe the general characteristics and functions of muscle tissue.

C. Anatomy of Skeletal Muscle Tissue

3. List and describe the connective tissue components of muscle tissue.
4. Discuss the important relationship of nerves and blood vessels to skeletal muscle.
5. Explain, in detail, the histology of skeletal muscle.

D. Contraction of Muscle and Types of Skeletal Muscle Fibers

6. Discuss the sliding-filament mechanism.
7. Differentiate between muscle tone, atrophy, and hypertrophy.
8. Contrast the structure and function of the three types of skeletal muscle fibers.

E. Cardiac Muscle Tissue

9. Describe the structure and function of cardiac muscle tissue.

F. Smooth Muscle Tissue

10. Describe the structure and function of smooth muscle tissue.
11. Compare the location and function of single-unit and multiunit muscle tissue.

G. Regeneration of Muscle Tissue

H. Developmental Anatomy of the Muscular System

I. Aging and Muscle Tissue

J. Applications to Health

12. Define fibromyalgia, muscular dystrophy, myasthenia gravis, and abnormal contraction.

K. Key Medical Terms Associated with the Muscular System

SCIENTIFIC TERMINOLOGY

Define the following root terms (prefix or suffix) associated with this chapter.

Prefix/Suffix	Meaning	Sample Word
-algia	pain	
apo-	from	
electro-	electricity	
endo-	within	
epi-	upon	
-graph	to write	
hyper-	above	
hypo-	below	
-itis	inflammation	
myo-	muscle	
-lemma	sheath	
neuron-	sinew	
-oma	tumor	
para-	beyond	
peri-	around	
sarco-	flesh	

A. Types of Muscle Tissue   (page 225)

A1.   Fill in the blanks for the questions below.

a.   Smooth muscle tissue is involved with internal processes related to the maintenance of the

internal environment. It is _____ in its appearance and

_____ in its control.

b.   _____ muscle tissue forms the bulk of the wall of the heart. It is

_____ at the microscopic level and its action is involuntary.

c.   Skeletal muscle is attached primarily to _____ and moves the skeleton. It

is _____ in appearance and _____ in its control.

B. Characteristic and Functions of Muscle Tissue (pages 225–226)

B1. Complete the questions below relating to muscle characteristics and functions.

a. _____ is the ability of muscle tissue to return to its original shape after contraction or extension.

b. The ability of muscle tissue to be stretched without damaging the tissue is called

_____.

c. The shortening and thickening of muscle is best described as _____.

d. It has been estimated that approximately _____ of all body heat is generated by muscle contractions.

e. List four kinds of movement of substances within the body produced by the contraction of muscles.

1. _____

2. _____

3. _____

4. _____

f. This function involves the contraction of skeletal muscle to hold the body in stationary

positions: _____ _____ _____.

g. _____ is the ability to receive and respond to stimuli.

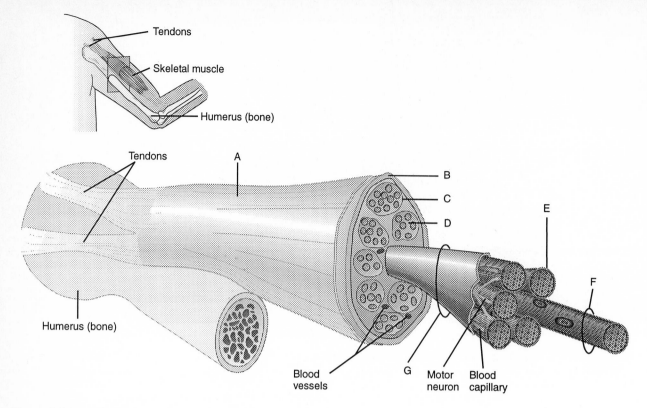

Tendons

Skeletal muscle

Humerus (bone)

Tendons

A

B

C

D

E

F

Humerus (bone)

Blood
vessels

G

Motor
neuron

Blood
capillary

Figure LG 9.1   Relation of connective tissue to skeletal muscle.

C. Anatomy of Skeletal Muscle Tissue   (pages 226–235)

C1.   Refer to Figure LG 9.1 and match the following names with the structures.

_____ deep fascia   _____ perimysium

_____ endomysium   _____ muscle fiber

_____ epimysium   _____ sarcolemma

_____ fascicle

C2.   Test your knowledge of connective tissue components and the nerve and blood supply of muscle.

1. endomysium	5. tendon
2. epimysium	6. tendon sheath
3. aponeurosis	7. one
4. perimysium	8. one or two

a.   _____ Certain tendons are enclosed in a fibrous connective tissue tube; found in the wrist and ankles.

b.   _____ The entire muscle is wrapped by this fibrous connective tissue.

c.   _____ Term applied to a sheet of tissue or broad, flat tendon attached to skin or another muscle or the coverings of bone.

d.   _____ A connective tissue cord that attaches a muscle to the periosteum of a bone.

e.   _____ A fibrous connective tissue that penetrates into the interior of each fascicle and surrounds the muscle fibers.

f.   _____ Bundles of muscle cells (fasciculi) are covered by this fibrous connective tissue.

g.   Generally, _____ artery and _____ veins accompany each nerve that penetrates a skeletal muscle.

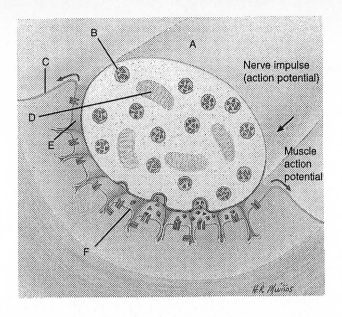

Enlarged view of the neuromuscular junction

Figure LG 9.2   Neuromuscular junction.

C3.   Test your knowledge of the neuromuscular junction and motor unit.

a.   Refer to Figure LG 9.2 and identify the following structures.

_____ mitochondrion   _____ synaptic cleft

_____ motor end plate   _____ synaptic end bulb

_____ sarcolemma   _____ synaptic vesicle

b.   The neurotransmitter released at neuromuscular junctions is _____.

c.   A neuron that stimulates muscle tissue is called a _____ neuron.

d.   A motor unit is defined as

e.   Muscles, such as the gastrocnemius, may have as many as _____ muscle fibers in each motor unit.

f.   As an axon nears its target skeletal muscle fiber, it branches into several

_____ _____.

g.   A typical neuromuscular junction has approximately *(10–20? 30–40?)* million acetylcholine receptors.

h.   Storage of the neurotransmitter substance occurs inside membrane-enclosed sacs called

_____ _____.

C4. Complete the following questions on muscle histology.

a. Muscle fibers are also referred to as _____.

b. Each muscle fiber is enveloped by a plasma membrane called the *(sarcoplasm? sarcolemma?)*.

c. Muscle cells possess a *(small? large?)* number of mitochondria.

d. Dilated end sacs of the sarcoplasmic reticulum are called _____.

    1. transverse tubules
    2. terminal cisterns
    3. a triad

e. The narrow *(H? M?)* zone in the center of each A band contains thick but not thin filaments.

f. A _____ consists of a transverse tubule and the terminal cisterns on either side of it at the A–I band.

g. The cylindrical structures, extending lengthwise, found in a muscle fiber are the *(myofibrils? myofilaments?)*.

h. Two protein filaments are found in muscle cells: the thin one contains the contractile protein _____ and the thick one contains the contractile protein

_____.

i. Sarcomeres are separated from one another by _____.

j. The dense (darker) area within a sarcomere is called the _____, while the

light area is called the _____.

k. The *(thin? thick?)* myofilament is anchored to the Z disc.

l. Each actin contains a _____ binding site.

m. Two other proteins are involved in the regulation of muscle contraction. They are

_____ and _____.

n. Myosin is anchored to the Z discs by a recently discovered _____

_____.

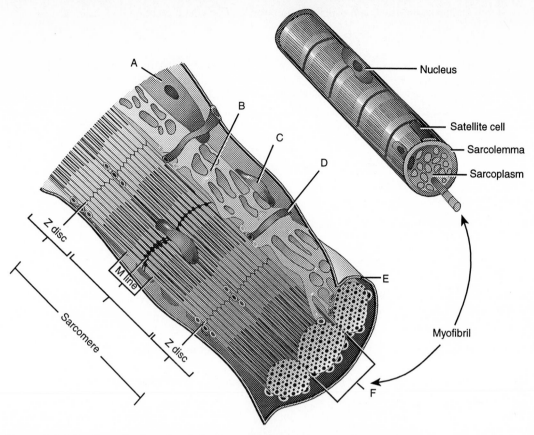

Figure LG 9.3 Organization of a skeletal muscle from gross to molecular levels.

C5. Refer to Figure LG 9.3 and answer the following questions.

a. Match the letters on the figure with the following structures; then color each structure. Be sure your color matches the color code oval.

O _____ mitochondrion   O _____ sarcolemma

O _____ myofibrils   O _____ sarcoplasmic reticulum

O _____ nucleus   O _____ transverse tubule

b. Label the I bands, A band, and H zone.

D. Contraction of Muscle and Types of Skeletal Muscle Fibers   (pages 235–238)

D1. Summarize the sliding-filament mechanism.

D2. Define muscle tone and explain its importance (for which function it is essential).

D3. Compare the three types of skeletal muscle fibers by completing the following table.

TYPES	Slow Fibers	Intermediate Fibers	Fast Fibers
Fatigability		Fatigue resistant	
Mitochondria	Many		
Myoglobin (amount)			Small amount
Capillaries		Many	
Split ATP	Slowly		
Diameter		Intermediate	

D4. The variation in the color of skeletal muscle fiber is dependent upon the concentration of

_____, a red-colored, oxygen-storing protein.

D5. Explain the relationship between the three types of skeletal muscle fibers and a motor unit.

D6. Briefly describe the effects of endurance training on fast glycolytic (type IIB) fibers.

D7. Complete this exercise about muscular atrophy and hypertrophy.

a. _____ _____ is a wasting away of muscles.

b. Bedridden individuals and people with casts may experience this type of atrophy:

_____ _____.

c. An increase in the diameter of muscle fibers due to the production of more myofibrils is

called _____ _____.

d. Anabolic steroids are responsible for increased muscle mass. They are similar to the male

sex hormone _____.

e. If the nerve supply to a muscle is cut, it will undergo _____ atrophy.

E. Cardiac Muscle Tissue (pages 238–240)

E1. Compare the structure and function of skeletal and cardiac muscle; then complete the table.

Characteristic	Skeletal Muscle	Cardiac Muscle
Number of nuclei (multinucleated/single)		
Number of mitochondria (more/less)		
Striated appearance (yes/no)		
Arrangement of fibers (parallel/branching)		
Nerve stimulation for contraction (yes/no)		
Length of refractory (long/short)		
Speed of contraction (fast/moderate)		

E2. Refer to Figure LG 9.4 and identify the following structures. After matching the structures color in each structure. Be sure to match the color code ovals.

O _____ desmosomes

O _____ gap junction

O _____ intercalated discs

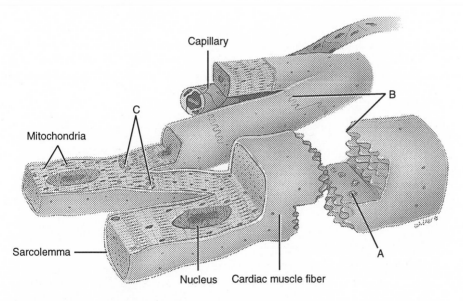

Figure LG 9.4 Histology of cardiac muscle tissue.

F. Smooth Muscle Tissue   (pages 240–241)

F1. Complete the table below of the principal characteristics of cardiac and smooth muscle tissue.

Characteristic	Cardiac Muscle	Smooth Muscle
Location		
Appearance		
Nervous control		
Sarcomeres		
T tubules		
Gap junctions		
Speed of contraction		

F2. Describe the function of the intermediate filaments and dense bodies with respect to smooth muscle cell contraction.

F3. Differentiate between visceral (single-unit) smooth muscle tissue and multiunit smooth muscle tissue.

F4. The decrease in tension which occurs in a smooth muscle fiber a minute or so after

initially contracting is referred to as the _____ _____.

G. Regeneration of Muscle Tissue   (page 241–242)

G1. Using the terms considerable, limited, and none, what is the capacity for regeneration in:

a. skeletal muscle—_____

b. cardiac muscle—_____

c. smooth muscle—_____

G2. What is a pericyte? Describe its function.

H. Developmental Anatomy of the Muscular System   (page 243)

H1. Somites differentiate into three regions: myotomes, dermatomes, and sclerotomes. What muscles form from each of these regions?

a. Except for a few muscles, all muscles of the body are derived from _____.

b. The first somites appear on the _____ day of embryologic development.

c. Eventually, _____ pairs of somites form by the _____ day.

I. Aging and Muscle Tissue   (page 243)

I1.   Discuss the effects of aging on muscle tissue.

J. Applications to Health (Refer to *Applications to Health* companion to answer Part J questions.)

J1.   Define fibromyalgia and myasthenia gravis.

a. Fibromyalgia

b. Myasthenia gravis

J2.   What are muscular dystrophies?

The most common form is _____. What protein is present in normal

muscle yet absent in persons with DMD? _____

J3.   Answer these questions pertaining to abnormal contractions of skeletal muscle.

a. A _____ refers to a painful spasmodic contraction.

b. An involuntary, brief twitch of a muscle visible under the skin is called a

_____.

c. The spasmodic twitching made involuntarily by muscles normally under voluntary control

is called a _____.

d. _____ is a rhythmic, involuntary, purposeless contraction of opposing
muscle groups.

e. _____ is similar to (b) above but is not visible under the skin.

ANSWERS TO SELECT QUESTIONS

A1. (a) Nonstriated, involuntary; (b) cardiac, striated, (c) bone, striated, voluntary.

B1. (a) Elasticity; (b) extensibility; (c) contractility (d) 85%; (e) 1—movement of food through gastrointestinal tract, 2—movement of blood within blood vessels, 3—movement of sperm and ova through the reproductive system, 4—movement of urine through the urinary system; (f) stabilizing body position; (g) excitability.

C1. A—Deep fascia, B—Epimysium, C—Perimysium, D—Endomysium, E—Sarcolemma, F—Muscle fiber, G—Fascicle.

C2. (a) 6; (b) 2; (c) 3; (d) 5; (e) 1; (f) 4; (g) 7, 8.

C3. (a) A—synaptic end bulb, B—synaptic vesicle, C—sarcolemma, D—mitochondrion; E—synaptic cleft, F—motor end plate; (b) acetylcholine; (c) motor; (e) 2,000; (f) axon terminals; (g) 30–40; (h) synaptic vesicles.

C4. (a) Myofibers; (b) sarcolemma; (c) large; (d) 2; (e) H; (f) triad; (g) myofibrils; (h) actin, myosin; (i) Z discs; (j) A band, I band; (k) thin; (l) myosin; (m) troponin, tropomyosin; (n) elastic myofilament.

C5. (a) A—nucleus, B—sarcoplasmic reticulum, C—mitochondrion, D—Transverse tubule, E—sarcolemma, F—myofibrils.

D7. (a) Muscle atrophy; (b) disuse atrophy; (c) muscular hypertrophy; (d) testosterone; (e) denervation.

D3.

Types	Slow Fibers	Intermediate Fibers	Fast Fibers
Fatigability	**Fatigue resistant**	Fatigue resistant	**Fatigable fibers**
Mitochondria	Many	**Many**	Few
Myoglobin (amount)	**Large**	**Large**	Small
Capillaries	**Many**	Many	Few
Split ATP	Slowly	**Rapid**	**Rapid**
Diameter	**Smallest**	Intermediate	**Largest**

D4. Myoglobin.

E1.

Characteristic	Skeletal Muscle	Cardiac Muscle
Number of nuclei (multinucleated/single)	**Multinucleated**	Single
Number of mitochondria (more/less)	**Less**	**More**
Striated appearance (yes/no)	**Yes**	**Yes**
Arrangement of fibers (parallel/branching)	**Parallel**	**Branching**
Nerve stimulation for contraction (yes/no)	**Yes**	**No**
Length of refractory (long/short)	**Short**	**Long**
Speed of contraction (fast/moderate)	**Fast**	**Moderate**

E2. A—gap junction, B—intercalated discs, C—desmosomes.

F1.

Characteristic	Cardiac Muscle	Smooth Muscle
Location	**Heart**	**Walls of hollow viscera**
Appearance	**Striated**	**Nonstriated**
Nervous control	**Involuntary**	**Involuntary**
Sarcomeres	**Yes**	**No**
T tubules	**Large**	**No**
Gap junctions	**Yes**	**Only in single-unit tissue**
Speed of contraction	**Moderate**	**Slow**

F4. Stress-relaxation response.

G1. (a) Limited; (b) none; (c) considerable.

G2. Stem cells found in association with the endothelium of blood capillaries and small veins from which smooth muscle fibers can arise.

H1. (a) mesoderm; (b) 20th; (c) 44, 30th.

J2. Duchenne muscular dystrophy; dystrophin.

J3. (a) Cramp; (b) fasciculation; (c) tic; (d) tremor; (e) fibrillation.

SELF QUIZ

Choose the one best answer to the following questions.

1. The structure responsible for calcium storage in a muscle is the

   A. transverse (T) tubules
   B. myofibril
   C. sarcoplasmic reticulum
   D. myofilaments
   E. sarcolemma

2. Which type of muscle tissue features striations, one nucleus, and only one transverse tubule per sarcomere?

   A. skeletal
   B. cardiac
   C. smooth
   D. none of the above are correct

3. The connective tissue sheath that surrounds bundles of muscle fibers is the

A. endomysium
B. epimysium
C. perimysium
D. superficial fascia
E. none of the above are correct

4. Approximately _____ % of body heat is generated by muscle contraction.

A. 50
B. 65
C. 75
D. 80
E. 85

5. A motor unit is defined as a

A. nerve and a muscle
B. single neuron and a single muscle fiber
C. neuron and the muscle fibers it supplies
D. single muscle fiber and the nerves that innervate it
E. muscle and the motor and sensory nerves that innervate it

6. Striations in skeletal muscle fiber are created by the

A. actin
B. myosin
C. troponin
D. tropomyosin
E. A and I bands

7. The junction of a transverse tubule and the terminal cisterns on either side is called a(n)

A. release channel
B. triad
C. sarcomere
D. A band
E. I band

8. All of the following are parts of thin filaments except

A. myosin
B. troponin
C. tropomyosin
D. actin
E. none of the above are correct

9. Dense bodies are associated with

A. cardiac muscle
B. type I fibers
C. type IIB fibers
D. smooth muscle
E. skeletal muscle

Answer (T) True or (F) False to the following questions.

10. _____ Fascia is one type of skeletal muscle tissue.

11. _____ A sarcomere lengthens during a muscle contraction.

12. _____ Skeletal muscle fibers are multinucleated.

13. _____ Myopathy refers to a pain in or associated with muscles.

14. _____ Acetylcholine is a neurotransmitter released at the myoneural junction.

15. _____ Cardiac muscle tissue has well-circumscribed myofibrils.

16. _____ Myomalacia refers to a softening of a muscle.

17. _____ The neuron and muscle cell are in direct contact at the motor end plate.

18. _____ Fast fibers can contract quickly and resist fatigue.

Fill in the blanks.

19. A _____ is a spasmodic twitching made involuntarily by muscles that are ordinarily under voluntary control.

20. The area between two Z discs is called the _____.

21. Each fiber in a network of the heart is separated from the next by an irregular transverse thickening of the sarcolemma called the _____ _____.

22. _____ _____ is a state of rigidity following death, due to the lack of ATP and the permanent binding of actin and myosin.

23. _____ is the reddish pigment, similar to hemoglobin, that is found in skeletal muscle fibers.

24. Upon removal of a cast from a healed broken arm, you might notice some _____ atrophy.

25. The light band in a skeletal muscle cell contains _____ filaments.

ANSWERS TO THE SELF QUIZ

1. C	10. F	19. tic
2. B	11. F	20. Sarcomere
3. C	12. T	21. Intercalated discs
4. E	13. F	22. Rigor mortis
5. C	14. T	23. Myoglobin
6. E	15. F	24. Disuse
7. B	16. T	25. Thin
8. A	17. F	
9. D	18. F	

The Muscular System

SYNOPSIS

In the last chapter your journey gave you a detailed examination of muscle tissue. **Muscle tissue** refers to all contractile tissue: skeletal, cardiac, and smooth. The **muscular system,** however, refers to the *skeletal* muscle system.

There are nearly 700 skeletal muscles that constitute the human muscular system. They receive their names based upon several types of characteristics such as size, shape, action, location, origin and insertion, number of origins, and the direction of the muscle fibers.

As your *fantastic voyage* through the body continues, you will learn how skeletal muscles produce movement, what an origin and insertion is, and how a lever system works. In addition, you will have the opportunity to study some of the principal skeletal muscles; their **origin**, **insertion**, **action**, and **innervation**.

TOPIC OUTLINE AND OBJECTIVES

A. How Skeletal Muscles Produce Movement

1. Define the terms origin, insertion, and gaster.
2. Describe the relationship between bones and skeletal muscles in providing movement.
3. Define a lever and fulcrum, and differentiate between the three types of levers as they relate to fulcrum, effort, and resistance.
4. Define and describe the types of skeletal muscle fiber arrangements and their correlation with the power and range of motion.
5. Define and explain the roles of the prime mover (agonist), antagonist, synergist, and fixator.

B. Naming Skeletal Muscles

6. Define the criteria utilized in naming skeletal muscles.

C. Principal Skeletal Muscles of the Head and Neck

D. Principal Skeletal Muscles That Act on the Abdominal Wall, Muscles Used in Breathing, Muscles of the Pelvic Floor and Perineum

E. Principal Skeletal Muscles That Move the Pectoral (Shoulder) Girdle and Upper Limb (Extremity)

F. Principal Skeletal Muscles That Move the Vertebral Column (Backbone)

G. Principal Skeletal Muscles That Move the Lower Limb (Extremity)

SCIENTIFIC TERMINOLOGY

Find an anatomical sample word for each prefix and suffix:

Prefix/Suffix	Meaning
agogos-	leader
anti-	against
bucc-	mouth
corrugo-	wrinkle
costa-	ribs
crani-	skull
dia-	across
epi-	over
ilio-	illium
labii-	lip
multi-	many
oculus-	eye
or-	mouth
orb-	circular
pectus-	breast
pubo-	pubis
rectus-	fiber parallel to midline
semi-	half
stylo-	stake or pole
sub-	under
supra-	above
syn-	together
-teres	round or long
thyro-	thyroid

A. How Skeletal Muscles Produce Movement (pages 258–263)

A1.   Define origin, insertion, and gaster.

a. Origin

b. Insertion

c. Gaster (belly)

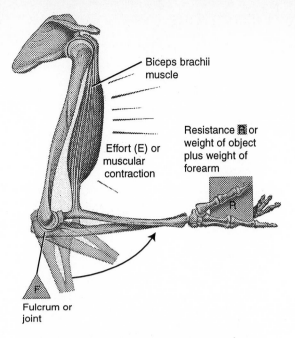

Biceps brachii
muscle

Resistance ☒ or
weight of object
plus weight of
forearm

Effort (E) or
muscular
contraction

Fulcrum or
joint

Movement of the forearm lifting a weight

Figure LG 10.1 Relationship of skeletal muscles to bones.

A2. Review Figure LG 10.1 and identify the following.

a. What forms the fulcrum? _____

b. Resistance is equal to the weight of the _____ plus the weight of the

_____.

c. The forearm is acting as a _____.

d. Effort is exerted by which muscle? _____ _____

e. If you stand on your toes, the weight of your body serves as the *(effort? fulcrum? resis-tance?)*, while the muscles of the calf provide the *(effort? fulcrum? resistance?)*. The

fulcrum would be the joint between phalanges and the _____.

This is an example of a _____-class lever.

A3. Match the following categories of levers with their proper descriptions (you may use the answers more than once).

F first-class levers
S second-class levers
T third-class levers

a. _____ Consist of the fulcrum at one end, the resistance at the opposite end, and the effort between them.

b. _____ Have the fulcrum between the effort and the resistance.

c. _____ Have the fulcrum at one end, the effort at the opposite end, and the resistance between them.

d. _____ An example is the head resting on the vertebral column.

e. _____ An example is flexing the forearm at the elbow.

f. _____ These are levers of strength; however, they sacrifice speed and range of motion (ROM).

A4. List the seven types of pattern arrangements of fascicles. (Review Table 10.1 in your textbook.)

a. _____

b. _____

c. _____

d. _____

e. _____

f. _____

g. _____

A5. Complete the following questions dealing with group actions.

a. If you were asked to extend your forearm (from a flexed position), the triceps brachii would

be the _____, and the biceps brachii would therefore be the _____.

b. A _____ serves to steady a movement, by preventing unwanted movements and helping the prime mover function more efficiently.

c. Synergistic muscles that stabilize the origin of the agonist so that the prime mover can act more efficiently are called _____.

A6. Answer these theoretical questions pertaining to group actions.

a. What would occur if a prime mover and an antagonist contracted at the same time?

b. What would be the net effect if the muscles on the posterior thigh (the hamstrings) did not function during contraction of the anterior thigh (the "quads")?

B. Naming Skeletal Muscles (pages 263–265)

B1. Using the terms given below, complete the following list. (Review Table 10.2 in your textbook.) Note: some questions have two answers.

1. direction of the fibers	5. shape
2. location	6. origin and insertion
3. size	7. action
4. number of origins	

a. _____ rectus abdominis

b. _____ sternocleidomastoid

c. _____ frontalis

d. _____ peroneus brevis

e. _____ trapezius

f. _____ quadriceps femoris

g. _____ adductor longus

h. _____ flexor carpi radialis

i. _____ transverse abdominis

j. _____ gluteus maximus

k. _____ biceps brachii

l. _____ stylohyoid

m. _____ tibialis anterior

n. _____ rhomboideus major

C. Principal Skeletal Muscles of the Head and Neck (pages 268–289)

C1. Test your knowledge of the muscles of facial expression. (See Exhibit 10.1.)

a. Contraction of the _____ _____ draws the angle of the mouth superiorly and laterally, as in smiling or laughing.

b. The _____ is responsible for pressing the cheek against the teeth or caving it in; it is a major cheek muscle.

c. To draw the outer part of the lower lip inferiorly and posteriorly, as in pouting, you would

contract the _____.

d. The orbicularis oculi has its origin on the _____ and inserts in the

_____.

e. The _____ _____ closes the lips, compresses the lips against the teeth, and shapes the lips during speech.

f. All of the muscles discussed in questions a–e are innervated by the

_____ (_____) nerve or cranial nerve.

g. If you fell asleep in class today, this muscle of the eyelid would have stopped functioning:

_____ _____ _____.

C2. Complete these questions on the muscles that move the eyeball. (See Exhibit 10.2.)

a. Label Figure LG 10.2, the extrinsic muscles of the eyeball.

_____ superior rectus   _____ medial rectus

_____ inferior rectus   _____ superior oblique

_____ lateral rectus   _____ inferior oblique

Figure LG 10.2 Extrinsic muscles of the eyeball.

b. Fill in the correct cranial nerve for each extrinsic eye muscle in the table below. (One is done for you.)

Muscle	Innervation
1. Superior rectus	
2. Inferior rectus	
3. Lateral rectus	
4. Medial rectus	Oculomotor, III
5. Superior oblique	
6. Inferior oblique	

C3. Answer these questions pertaining to muscles that move the mandible (lower jaw). (See Exhibit 10.3.)

a. The _____ has its origin on the maxilla and zygomatic arch and inserts into the angle and ramus of the mandible.

b. The temporalis originates on the _____ and _____

bones and inserts into the _____ and _____ of the

mandible.

c. The muscles in questions a and b *(depress? elevate?)* the mandible.

d. The muscles that move the lower jaw are innervated by cranial nerve *(IV? V? VI?)*, which is

called the _____ nerve.

e. Refer to Figure LG 10.3a, and review the muscles of the head and neck.

C4. Match the groups of muscles with the descriptions below. (See Exhibits 10.4, 10.5, 10.6, and 10.7.)

> 1. genioglossus, styloglossus, hypoglossus
> 2. cricothyroid, thyrocrytenoid
> 3. stylohyoid, mylohyoid, geniohyoid
> 4. omohyoid, sternohyoid, thyrohyoid
> 5. lateral cricoarytenoid, arytenoid

a. _____ Function to close the rima glottidis.

b. _____ Are involved in movement of the tongue. These muscles are innervated by the hypoglossal (XII) nerve.

c. _____ Contraction of these infrahyoid muscles depresses the hyoid bone.

d. _____ These suprahyoid muscles will elevate the hyoid when contracted.

e. _____ These muscles elongate and place tension on the vocal folds or shorten and relax the vocal folds, respectively.

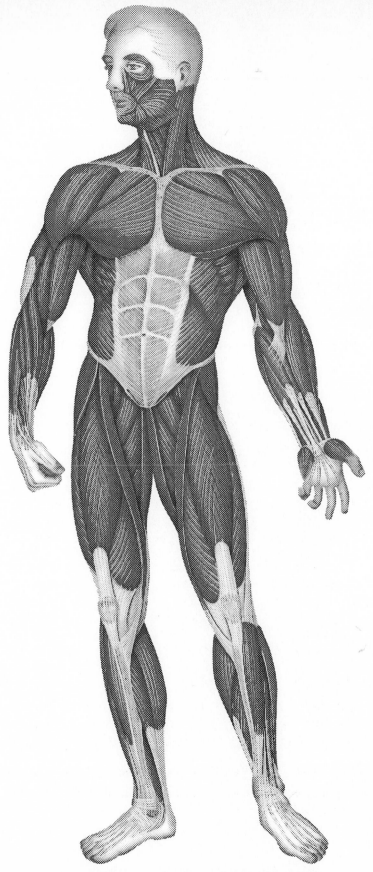

Figure LG 10.3a Principal superficial skeletar muscles.

Figure continues

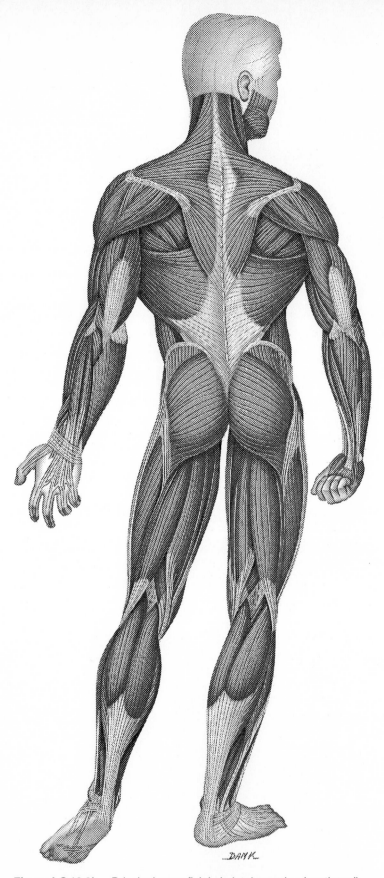

Figure LG 10.3b Principal superficial skeletal muscles (continued).

C5. Answer the following questions about the muscles that move the head. (See Exhibit 10.8.)

a. Contraction of your right sternocleidomastoid would rotate your head to the *(right? left?)*.

b. If you were to contract the left and right sternocleidomastoid muscles simultaneously, what

action would occur to the head? _____

c. The splenius capitis is innervated by the _____ _____
of the middle and inferior cervical nerves.

d. The _____ _____ extends the head and rotates the
face toward the opposite side as the contracting muscle.

e. The occipital bone and mastoid process of the temporal bone are the insertion points for

which muscle? _____ _____

D. Principal Skeletal Muscles That Act on the Abdominal Wall, Muscles Used in Breathing, Muscles of the Pelvic Floor and Perineum (pages 290–299)

D1. There are four pairs of muscles that form the anterolateral abdominal wall. Answer the
questions below pertaining to these muscles. (See Exhibit 10.9.)

a. The fibers of the _____ _____ run parallel to the

midline, while the fibers of the _____ _____ run
perpendicular.

b. Contraction of one of the external oblique muscles bends the vertebral column *(medially? laterally?)*.

c. The _____ _____ originates on the iliac crest, inguinal
ligament, and thoracolumbar fascia and inserts on the cartilage of the last *(2 or 3? 3 or 4?)*
ribs.

d. Contraction of the bilateral rectus abdominis would *(flex? extend?)* the vertebral column.

e. Refer to Figure LG 10.3a, and review the muscles of the abdominal wall.

D2. Answer these questions about the muscles of breathing. (See Exhibit 10.10.)

a. The muscle that forms the line of separation between the thoracic and abdominopelvic

cavity is the _____.

b. The internal intercostals originate on the *(superior? inferior?)* border of the rib *(below? above?)*.

c. The external intercostals *(depress? elevate?)* the ribs during inspiration.

d. Contraction of the diaphragm causes a(n) *(decrease? increase?)* in the volume of the
thoracic cavity.

e. The _____ nerve innervates the diaphragm.

D3. Answer the following questions about the muscles of the pelvic floor. (See Exhibits 10.11 and 10.12.)

a. Which structures compose the pelvic diaphragm? What is its function?

b. Name the two parts that form the levator ani.

　1. _____

　2. _____

D4. Which structures form the borders of the perineum?

a. Anteriorly _____ _____

b. Laterally _____ _____

c. Posteriorly _____

D5. List the muscles that form the pelvic floor.

E. Principal Skeletal Muscles That Move the Pectoral (Shoulder) Girdle and Upper Limb (Extremity)   (pages 300–324)

E1. Match the muscles listed below to their descriptions. (See Exhibit 10.13.)

LS levator scapulae	SA serratus anterior
PM pectoralis minor	SU subclavius
RMM rhomboideus major and minor	TR trapezius

a. _____ Adducts the scapula and slightly rotates it downward.

b. _____ Depresses the clavicle anteriorly and stabilizes the pectoral girdle.

c. _____ Abducts and rotates the scapula upward and elevates ribs when the scapula is fixed.

d. _____ Adducts the scapula, rotates the scapula, elevates or depresses the scapula, and extends the head.

e. _____ Elevates the scapula and slightly rotates it downward.

f. _____ Depresses and abducts the scapula and elevates the third through fifth ribs during forced inhalation.

E2. Complete the following table pertaining to muscles that move the arm (humerus). (See Exhibit 10.14.)

Muscle	Origin	Insertion
a. Pectoralis major		
b. Latissimus dorsi		
c. Deltoid		
d. Supraspinatus		
e. Infraspinatus		
f. Teres major		

Refer to Exhibit 10.15 to answer questions E3 and E4.

E3. The muscles that move the forearm (radius and ulna) are divided into

_____ and extensors.

E4. Complete the table below, pertaining to the muscles that are involved in forearm movement.

Muscle	Origin	Insertion
a. Biceps brachii		
b. Triceps brachii		
c. Pronator teres		

E5. Complete the following questions about the muscles that move the wrist, hand, and fingers. (See Exhibit 10.16.)

a. The _____ compartment muscles function as flexors while the

_____ compartment muscles function as extensors.

b. The _____ _____ _____ flexes the
distal phalanx of the thumb.

c. Which muscle flexes and abducts the wrist and originates at the medial epicondyle of the

humerus? _____ _____ _____

d. Which extensor muscle inserts on the second metatarsal? _____

_____ _____ _____

e. Which three muscles of the hand would a hitchhiker be contracting?

1. _____

2. _____

3. _____

E6. Refer to Figure LG 10.3a, and review the muscles that move the pectoral girdle and upper limb. Perform this exercise for Figure LG 10.3b.

F. Principal Skeletal Muscles That Move the Vertebral Column (Backbone) (pages 325–330)

Refer to Exhibit 10.18 to answer questions F1–F4.

F1. The splenius muscles arise from the midline of the body and run

_____ and _____ to their insertion.

F2. The three groupings that comprise the erector spinae are the

a. _____

b. _____

c. _____

F3. The erector spinae function to *(flex? extend?)* the vertebral column.

F4. The scalenes _____ and _____ the head and assist

in deep _____.

F5. Why would a sports physician recommend sit-ups as one exercise that aids in the treatment of lower back pain?

G. Principal Skeletal Muscles That Move the Lower Limb (Extremity) (pages 331–352)

G1. Match the following muscles to their description. (See Exhibit 10.19.)

AL	adductor longus	PR piriformis
GM	gluteus maximus	PS psoas major
GMD	gluteus medius	TFL tensor fasciae latae

a. _____ Adducts, medially rotates, and flexes the thigh.

b. _____ Flexes and rotates thigh laterally; flexes the trunk on the hip.

c. _____ Abducts and rotates thigh medially.

d. _____ Rotates thigh laterally and abducts it.

e. _____ Extends and rotates thigh medially.

f. _____ Flexes and abducts the thigh.

G2. Complete the following table about the quadriceps group. (See Exhibit 10.20 to answer questions G2 and G3.)

Muscle	Origin	Insertion
a. Rectus femoris		
b. Vastus lateralis		
c. Vastus medialis		
d. Vastus intermedius		

G3. The hamstrings are composed of the _____ _____,

_____, and _____. Each performs the same function

of _____ the leg and _____ the thigh.

G4. The _____ is the longest muscle in the body, originating on the anterior superior iliac spine and inserting into the medial surface of the body of the tibia.

Refer to Exhibit 10.21 to complete the following questions.

G5. The lateral (peroneal) compartment contains two muscles that _____

flex and _____ the foot.

G6. The _____ anterior dorsiflexes and inverts the foot.

G7. (a) Name the three superficial muscles that form the posterior compartment of the lower leg and indicate their (b) insertion, (c) function, and (d) innervation.

a. 1. _____

    2. _____

    3. _____

b. _____

c. _____

d. _____

G8. Identify the following functions and muscle action (some questions require two answers).

DF	dorsiflexion
PF	plantar flexion
IN	inversion
EV	eversion

a. _____ Walking on the sides of your feet.   e. _____ Tibialis anterior

b. _____ Standing on tiptoe.   f. _____ Peroneus longus

c. _____ Walking on your heels.   g. _____ Tibialis posterior

d. _____ Moving your sole laterally.   h. _____ Extensor Hallucis longus

G9. Refer to Figure LG 10.3a, and review the muscles that move the lower limb. Repeat this exercise for Figure LG 10.3b.

ANSWERS TO SELECT QUESTIONS

A2. (a) Elbow; (b) object, forearm; (c) lever; (d) biceps brachii; (e) resistance, effort, metatarsals, second.

A3. (a) T; (b) F; (c) S; (d) F; (e) T; (f) S.

A4. (a) Parallel; (b) fusiform; (c) unipennate; (d) bipennate; (e) multipennate; (f) circular, (g) triangular.

A5. (a) Agonist, antagonist; (b) synergist; (c) fixators.

A6. (a) There would be NO movement. The muscle actions would counter each other. (b) The reflex (contraction) would be greatly exaggerated.

B1. (a) 1, 2; (b) 6; (c) 2; (d) 3; (e) 5; (f) 2, 4; (g) 2, 7; (h) 2, 7; (i) 1, 2; (j) 3; (k) 4; (l) 6; (m) 2; (n) 3, 5.

C1. (a) Zygomaticus major; (b) buccinator; (c) platysma; (d) medial wall of orbit, circular path around the orbit; (e) orbicularis oris; (f) facial; VII; (g) levator palpebrae superioris.

C2. (a) A—inferior oblique; B—superior oblique; C—medial rectus; D—lateral rectus; E—inferior rectus; F—superior rectus.

(b)

Muscle	Innervation
1. Superior rectus	Oculomotor, III
2. Inferior rectus	Oculomotor, III
3. Lateral rectus	Abducens, VI
4. Medial rectus	Oculomotor, III
5. Superior oblique	Trochlear, IV
6. Inferior oblique	Oculomotor, III

C3. (a) Masseter; (b) temporal frontal, coronoid process, ramos; (c) elevate; (d) V, trigeminal.

C4. (a) 5; (b) 1; (c) 4; (d) 3; (e) 2.

C5. (a) Left; (b) flexion; (c) dorsal rami; (d) semispinalis capitis; (e) splenius capitis.

D1. (a) Rectus abdominis, transverus abdominis; (b) laterally; (c) internal oblique, 3 or 4; (d) flex.

D2. (a) Diaphragm; (b) superior, below; (c) elevate; (d) increase; (e) phrenic.

D3. (a) Muscles of the pelvic floor together with the fascia covering their external and internal surfaces. They support the pelvic viscera. (b) 1. pubococcygeus; 2. iliococcygeus.

D4. (a) Pubic symphysis; (b) ischial tuberosities; (c) coccyx.

D5. Levator ani (pubococcygeus and iliococcygeus), coccygeus.

E1. (a) RMM; (b) SU; (c) SA; (d) TR; (e) LS; (f) PM.

E2.

Muscle	Origin	Insertion
a. Pectoralis major	Clavicle, sternum, cartilage of second to sixth ribs	Greater tubercle and intertubercular sulcus of humerus
b. Latissimus dorsi	Spines of inferior six thoracic vertebrae, lumbar vertebrae, crests of sacrum and ilium, inferior four ribs	Intertubercular sulcus of humerus
c. Deltoid	Acromial extremity of clavicle and acromion and spine of scapula	Deltoid tuberosity of humerus
d. Supraspinatus	Supraspinatus fossa of scapula	Greater tubercle of humerus
e. Infraspinatus	Infraspinatus fossa of scapula	Greater tubercle of humerus
f. Teres major	Inferior angle of scapula	Intertubercular sulcus of humerus

E3. Flexors.

E4.

Muscle	Origin	Insertion
a. Biceps brachii	Long head originates from tubercle above glenoid cavity; short head originates from coracoid process of scapula.	Radial tuberosity and bicipital aponeurosis
b. Triceps brachii	Long head originates from a projection inferior to glenoid cavity of scapula; lateral head originates from lateral and posterior surface of humerus superior to radial groove; medial head originates from entire posterior surface of humerus inferior to a groove for the radial nerve.	Olecranon of ulna
c. Pronator teres	Medial epicondyle of humerus and coronoid process of ulna.	Midlateral surface of radius

E5. (a) Anterior, posterior; (b) flexor pollicis longus; (c) flexor carpi ulnaris; (d) extensor carpi radialis longus; (e) 1. abductor pollicis longus, 2. extensor pollicis brevis, 3. extensor pollicis longus.

F1. Laterally, superiorly.

F2. (a) Iliocostalis; (b) longissimus; (c) spinalis.

F3. Extend.

F4. Flex, rotate, inspiration.

F5. Compresses the abdomen, thereby taking pressure and stress off the lumbar spine.

G1. (a) AL; (b) PS; (c) GMD; (d) PR; (e) GM; (f) TFL.

G2.

Muscle	Origin	Insertion
a. Rectus femoris	Anterior inferior iliac spine	Patella via quadriceps tendon and the tibial tuberosity via the patella ligaments
b. Vastus lateralis	Greater trochanter and linea aspera of femur	Patella via quadriceps tendon and the tibial tuberosity via the patella ligaments
c. Vastus medialis	Linea aspera of femur	Patella via quadriceps tendon and the tibial tuberosity via the patella ligaments
d. Vastus intermedius	Anterior and lateral surfaces of body of femur	Patella via quadriceps tendon and the tibial tuberosity via the patella ligaments

G3. Biceps femoris, semimembranosus, semitendinosus, flexing, extending.

G4. Sartorius.

G5. Plantar, evert.

G6. Tibialis.

G7. (a) 1. Gastrocnemius, 2. soleus, 3. plantaris; (b) calcaneus by way of the calcaneal or Achilles tendon; (c) plantar flexes the foot; (d) tibial nerve.

G8. (a) IN; (b) PF; (c) DF; (d) EV; (e) DF, IN; (f) PF, EV; (g) PF, IN; (h) DF, EV.

SELF QUIZ

Choose the one best answer to the following questions.

1. All of these muscles are located in the upper extremity except

   A. biceps brachii
   B. flexor carpi ulnaris
   C. pronator teres
   D. rhomboideus major
   E. brachioradialis

2. All these muscles are located on the posterior except

   A. trapezius
   B. quadratus lumborum
   C. sacrospinalis
   D. teres minor
   E. rectus abdominis

3. Which of the following muscles does NOT belong in the group?

   A. pectoralis major
   B. tibialis anterior
   C. gastrocnemius
   D. soleus
   E. all of the above belong in the group

4. The muscle responsible for extension of the thigh is the

   A. extensor pollicis longus
   B. gluteus maximus
   C. adductor longus
   D. tensor fasciae latae
   E. brachioradialis

5. The muscle involved in flexion of the head is the

   A. trapezius
   B. sternocleidomastoid
   C. multifidus
   D. semispinalis capitis
   E. levator scapulae

6. For proper chewing function, this cranial nerve must be functional.

   A. III
   B. IV
   C. V
   D. VI
   E. I

7. The portion of a muscle attached to a movable bone is called the

   A. gaster
   B. aponeurosis
   C. origin
   D. insertion
   E. none of the above answers is correct

8. Which muscle name is based upon shape?

   A. sternocleido-mastoid
   B. trapezius
   C. rectus abdominis
   D. deltoid
   E. B and D

9. The primary muscle responsible for a desired action (movement) is known as the

   A. synergist
   B. antagonist
   C. agonist
   D. fixator
   E. none of the above answers is correct

10. The longest muscle in the body is the

   A. rectus abdominis
   B. sartorius
   C. triceps brachii
   D. gastrocnemius
   E. biceps brachii

Answer (T) True or (F) False to the following questions.

11. _____ The triceps brachii is an antagonist to the biceps brachii.

12. _____ The name adductor longus is based on action only.

13. _____ Both the masseter and temporalis muscles are used for chewing.

14. _____ The corrugator supercilli draws the eyebrow inferiorly as in frowning.

15. _____ Standing on "tiptoe" involves contraction of the gastrocnemius.

16. _____ The trapezius muscle functions to flex the head.

17. _____ The extensors of the wrist are located on the posterior side of the forearm.

18. _____ The flexor pollicis brevis flexes the great toe.

19. _____ The origin of a muscle ordinarily attaches it to a stable, nonmoving bone.

Fill in the blanks.

20. Someone who is proficient at cheating on tests has developed this muscle of eye movement.

    _____ _____

21. The _____ aids in breathing and separates the thoracic and abdominopelvic cavities.

22. The fibers of the rectus abdominis run _____ to the midline of the body.

23. A muscle that increases the angle at a joint would be classified as a(n) _____.

24. The biceps brachii is classified as a _____.

25. The peroneus longus and peroneus brevis are located in the _____ compartment of the leg.

ANSWERS TO THE SELF QUIZ

1. D	10. B	19. T
2. E	11. T	20. Lateral rectus
3. A	12. F	21. Diaphragm
4. B	13. T	22. Parallel
5. B	14. T	23. Extensor
6. C	15. T	24. Flexor
7. D	16. F	25. Lateral
8. E	17. T	
9. C	18. F	

Surface Anatomy

CHAPTER
11

SYNOPSIS

Many of you have aspirations of entering the health profession. As future doctors, dentists, nurses, and other allied health professionals, none of you would argue that a thorough knowledge of the internal working of the body is an essential component of your education. In order to diagnose and render treatment for any disease or disorder, it is imperative that you understand the anatomy and physiology of the body.

One area often overlooked in the study of human anatomy is **surface anatomy.** Surface anatomy, as defined in Chapter 1, is the study of the **morphology (form)** and **marking of the surface of the body**. A knowledge of surface anatomy, when used with visual and tactile palpation, will help you identify specific superficial structures.

The importance of surface anatomy to health care professionals is seen in such activities as auscultation of the heart and lungs, taking a pulse, palpating the abdomen, drawing blood, and administering injections.

To introduce you to surface anatomy, some key features of each region will be discussed.

TOPIC OUTLINE AND OBJECTIVES

A. Introduction

1. Define surface anatomy and identify the principal regions of the human body.

B. Head

2. Describe the surface anatomy features of the head.

C. Neck

3. Describe the surface anatomy traits of the neck.

D. Trunk

4. Describe the surface anatomy features of the trunk.

E. Upper Limb (Extremity)

5. Describe the surface anatomy traits of the upper limb (extremity).

F. Lower Limb (Extremity)

6. Describe the surface anatomy features of the lower limb (extremity).

A. Introduction   (page 360)

A1. Define surface anatomy and palpation.

a. Surface anatomy

b. Palpation

B. Head   (pages 360–365)

Refer to Figure 11.1 in your textbook and to the chapter section on the eyes, ears, nose, and lips to answer the following questions.

B1. Match the following surface anatomy region terms to their descriptions.

1. orbital		6. occipital	
2. nasal		7. auricular	
3. buccal		8. parietal	
4. oral		9. temporal	
5. mental		10. frontal	

a._____ side of skull

b._____ nose

c._____ base of skull

d._____ cheek

e._____ front of skull

f. _____ ear

g. _____ anterior portion of mandible

h. _____ crown of skull

i. _____ includes eyeball and eyelids

j. _____ mouth

B2. The head is divided into two regions: the _____ and the

_____.

B3. Answer these questions pertaining to the eyes, ears, nose, and lips.

a. The *(tragus? antitragus?)* is the cartilaginous projection anterior to the external auditory canal.

b. The portion of the external ear not contained in the head is the *(tympanic membrane? auricle?)*.

c. The *(external naris? dorsum nasi?)* is the external opening into the nose.

d. The *(palpebrae? conjunctiva?)* is the membrane that covers the exposed surface of the eyeball and lines eyelids.

e. The "white" of the eye refers to the *(sclera? iris?)*.

f. The *(medial? lateral?)* commissure is the site of union of the upper and lower eyelids, away from the nose.

g. The circular pigmented muscular structure behind the cornea is called the *(pupil? iris?)*.

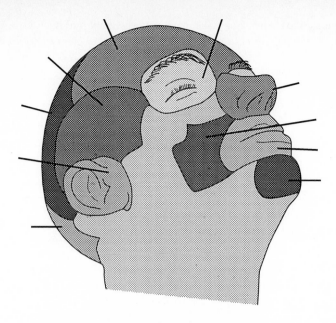

Figure LG 11.1   Regions of the head.

B4.   Refer to Figure LG 11.1 and label the regions of the head (Note: refer to the photograph in Figure 11.1 in your textbook).

C. Neck   (pages 365–368)

Refer to Figure 11.6 and the chapter section on the neck to answer the following questions.

C1.   The anterior triangle is bordered superiorly by the _____, inferiorly by

the _____, medially by the _____

_____, and laterally by the anterior border of the

_____ muscle.

C2.   The _____ pulse may be detected by palpating superior to the larynx and just anterior to the sternocleidomastoid.

C3.   The Adam's apple is in actuality the _____ _____.

C4.   A prominent vein that runs along the lateral surface of the neck is the

_____ _____ _____.

C5.   Describe the borders of the posterior triangle of the neck.

D. Trunk   (pages 368–372)

Refer to Figures 11.7, 11.8, and 11.9 and to the chapter section on the trunk to answer the following questions.

D1.   The trunk is divided into the _____, _____,

_____ and _____.

D2. You are an emergency room physician when a patient is brought into the ER complaining of right lower quadrant pain. Upon examination you apply finger pressure to McBurney's

point, which produces point tenderness. What is your diagnosis? _____

D3. Match the vertebral spine to the description below.

C7	T7
T3	L4

a. _____ The supracristal line passes through this spine.

c. _____ Found opposite the inferior angle of the scapula.

b. _____ Prominent at the base of the neck

d. _____ Is at the same level as the spine of the scapula.

D4. What forms the triangle of auscultation, and why is it important?

D5. Between the medial ends of the clavicles is a depression on the superior surface of the

sternum called the _____ _____. What structure can

be palpated there? _____

D6. Why is the fifth left intercostal space of importance?

D7. What surface landmark is formed by the junction between the manubrium and body of the sternum?

D8. Define the following terms.

a. Linea alba

b. Posterior axillary fold

c. Anterior axillary fold

d. Xiphoid process

E. Upper Limb (Extremity)   (pages 372–378)

Refer to Figures 11.10–11.13 and to the chapter section on the upper limb (extremity) to answer the following questions.

E1. The slight elevation at the lateral end of the clavicle is the _____ joint.

E2. The rounded prominence of the shoulder is formed by the *(supraspinatus? deltoid?)* muscle.

E3. What is the cubital fossa? Name the artery that passes through it and its importance.

E4. The _____ _____ is the lateral projection at the distal end of the humerus.

E5. The radial artery can be palpated just medial to the _____

_____ of the radius.

E6. Describe the "anatomical snuffbox." Which structures can be palpated there?

E7. Match the following terms to the anatomical region.

A antebrachium	C carpus
B brachium	M manus

a._____ wrist   c. _____ arm

b._____ hand   d. _____ forearm

E8. The ulnar *(nerve? artery? vein?)* can be palpated in a groove behind the medial epicondyle.

E9. The styloid process of the *(ulna? radius?)* is one of the most prominent landmarks at the medial side of the wrist.

E10. The common name used to describe the second through fifth metacarpal heads is the

"_____."

E11. The rounded contour formed by the thumb muscles is the *(thenar? hypothenar?)* eminence.

F. Lower Limb (Extremity)   (pages 378–382)

Refer to Figures 11.14–11.16 and the chapter section on the lower limb (extremity) to answer the following questions.

F1.   Match the terms below with the correct description.

CT	calcaneal tendon	LM	lateral malleolus
DVA	dorsal venous arch	MM	medial malleolus
GC	gluteal cleft	PF	popliteal fossa
GF	gluteal fold	PL	patellar ligament
IT	ischial tuberosity	TT	tibial tuberosity

a. _____ Superficial veins of foot that unite to form small and great saphenous veins.

b. _____ Runs from the gastrocnemius and soleus muscles and inserts into the calcaneus.

c. _____ Continuation of the quadriceps femoris tendon inferior to the patella.

d. _____ Bears weight of body when seated.

e. _____ Inferior limit of buttock, formed by inferior margin of gluteus maximus muscle.

f. _____ Projection of distal end of the fibula.

g. _____ Bony prominence of tibia into which patellar ligament inserts.

h. _____ Diamond-shaped space on posterior aspect of knee, visible when knee is flexed.

i. _____ Depression along midline that separates the buttocks.

j. _____ Projection of distal end of the tibia.

F2.   Match the alternative names.

CR	crus
FM	femoral
GL	gluteal
GN	genu
TR	tarsus

a. _____ buttocks   d. _____ knee

b. _____ ankle   e. _____ leg

c. _____ thigh

F3.   What structures border the femoral triangle? Name the structures contained in the femoral triangle.

F4.   The _____ _____ vein is the longest vein in the body.

ANSWERS TO SELECT QUESTIONS

B1. (a) 9; (b) 2; (c) 6; (d) 3; (e) 10; (f) 7; (g) 5; (h) 8; (i) 1; (j) 4.

B2. Cranium, face.

B3. (a) Tragus; (b) auricle; (c) external naris; (d) conjunctiva; (e) sclera; (f) lateral; (g) iris.

C1. Mandible, sternum, cervical midline, sternocleidomastoid.

C2. Carotid (neck).

C3. Thyroid cartilage.

C4. External jugular vein.

D1. Back, chest, abdomen, pelvis.

D2. Acute appendicitis.

D3. (a) L4; (b) C7; (c) T7; (d) T3.

D4. Latissimus dorsi, trapezius, vertebral border of scapula. Permits the clear auscultation of the respiratory sounds.

D5. Suprasternal notch, trachea.

D6. Auscultation of the apex beat of the heart.

D7. Sternal angle.

E1. Acromioclavicular.

E2. Deltoid.

E3. A triangular space on the anterior surface of the elbow. Brachial artery; common site for the taking of blood pressure.

E4. Lateral epicondyle.

E5. Styloid process.

E7. (a) C; (b) M; (c) B; (d) A.

E8. Nerve.

E9. Ulna.

E10. Knuckles.

E11. Thenar.

F1. (a) DVA; (b) CT; (c) PL; (d) IT; (e) GF; (f) LM; (g) TT; (h) PF; (i) GC; (j) MM.

F2. (a) GL; (b) TR; (c) FM; (d) GN; (e) CR.

F4. Great saphenous.

SELF QUIZ

Choose the one best answer to the following questions.

1. Which of the following structures is used as a landmark for a tracheostomy?

   A. hyoid bone
   B. thyroid cartilage
   C. cricoid
   D. sternocleidomastoid
   E. suprasternal notch

2. Which area of the head is associated with the term auricular?

   A. eyes
   B. cheeks
   C. base of the skull
   D. ears
   E. tongue

3. The dorsalis pedis pulse is located just lateral to the tendon of the _____ muscle.

   A. extensor hallucis longus
   B. extensor digitorum longus
   C. extensor digitorum brevis
   D. flexor hallucis brevis
   E. extensor sartorius

4. Which muscle borders the femoral triangle medially?

   A. rectus femoris
   B. adductor longus
   C. biceps femoris
   D. vastus lateralis
   E. vastus intermedius

5. What structure can be palpated in the groove behind the medial epicondyle of the humerus?

   A. humeral artery
   B. brachial artery
   C. brachial vein
   D. ulnar nerve
   E. radial nerve

6. Which of the following muscles is a frequent site for intramuscular injections?

   A. deltoid
   B. biceps brachii
   C. trapezius
   D. brachioradialis
   E. triceps brachii

7. What feature can be palpated along the anterior border of the sternocleidomastoid?

   A. jugular vein
   B. carotid pulse
   C. vertebral arteries
   D. subclavian arteries
   E. arch of the aorta

8. The spine of scapula is located at the level of the _____ vertebral spine.

   A. T3
   B. T4
   C. T5
   D. T6
   E. T7

9. The umbilicus is approximately on a level with the

    A. L3–L4 intervertebral disc
    B. L4–L5 intervertebral disc
    C. aortic bifurcation
    D. A and C
    E. none of the above answers are correct

10. The most frequent injection site of insulin in a diabetic is the

    A. biceps femoris   D. rectus femoris
    B. vastus lateralis   E. rectus abdominis
    C. vastus medialis

Answer (T) True or (F) False to the following questions.

11. _____ The radial pulse is located just lateral to the styloid process of the radius.

12. _____ The proximal end of the metacarpals form the "knuckles."

13. _____ The most prominent muscle of the anterior leg (crus) is the gastrocnemius.

14. _____ The triangle of auscultation is formed by the latissimus dorsi muscle, trapezius muscles, and the vertebral border of scapula.

15. _____ The most prominent spine of the cervical vertebrae is C6.

16. _____ The apex of the heart can be heard in the left fourth intercostal space, just medial to the left midclavicular line.

17. _____ The longest vein in the body is the femoral vein.

18. _____ The ischial tuberosity is the point of the ischium that bears our weight when seated.

19. _____ The vein most frequently used for venipuncture lies in the popliteal fossa.

20. _____ The inferior angle of the scapula is located opposite the spine of the T7 vertebra.

ANSWERS TO THE SELF QUIZ

1. C	8. A	15. F
2. D	9. D	16. F
3. A	10. B	17. F
4. B	11. F	18. T
5. D	12. F	19. F
6. A	13. F	20. T
7. B	14. T	

The Cardiovascular System: Blood

CHAPTER

12

SYNOPSIS

The surgeons and scientists who were miniaturized and injected into a human body in the movie *Fantastic Voyage* used the cardiovascular system as their highway through the body.

The function of the **cardiovascular system,** which consists of the **blood, heart,** and **blood vessels**, is to transport blood throughout the body. The blood is the medium that delivers oxygen, nutrients, and hormones to the body while removing carbon dioxide and wastes from the cells. Regulating body temperature and pH, preventing hemorrhage, and combating disease are also functions of the blood.

As the blood courses through the circulatory system, it picks up oxygen in the lungs, nutrients in the gastrointestinal tract, and hormones from the endocrine glands. Upon reaching the microscopic capillary beds (located throughout the body), materials in the blood diffuse into the interstitial fluid. **Interstitial fluid** (also known as **intercellular** or **tissue fluid**) bathes the cells of our bodies, allowing the exchange of nutrients and wastes to take place.

TOPIC OUTLINE AND OBJECTIVES

A. Functions of Blood

1. List and describe the three functions of blood.

B. Physical Characteristics of Blood

2. Describe the physical characteristics of blood.

C. Components of Blood

3. Describe the components of blood, formed elements, and plasma.
4. List the formed elements.
5. Describe the chemical composition and substances in the plasma.

D. Formation of Blood Cells

E. Erythrocytes (Red Blood Cells)

6. Discuss the origin, structure, function, life span, and number of the erythrocytes.
7. Explain the blood group system.

F. Leukocytes (White Blood Cells)

8. Describe the origin, histology, functions, life span, numbers, and production of the five types of leukocytes.

G. Platelets

9. Describe the origin, structure, function, life span, and number of the platelets.

H. Applications to Health

10. Contrast the causes and effects of anemia, infectious mononucleosis (IM), and leukemia.

I. Key Medical Terms Associated with Blood

SCIENTIFIC TERMINOLOGY

Find an anatomical sample word for each prefix and suffix:

Prefix/Suffix	Meaning	Sample Word
cardio-	heart	
cyano-	blue	
-cyte	cell	
erythros-	red	
hemo-	blood	
leuko-	white	
macro-	large	
phlebo-	vein	
-poiem	to make	
thrombo-	clot	
-vascular	blood vessel	
veno-	vein	

A. Functions of Blood   (page 377)

A1. Name the three components of the cardiovascular system.

a. _____

b. _____

c. _____

A2. The fluid that bathes differentiated body cells is called _____

_____.

A3. Complete the following statements pertaining to the function of blood.

a. It transports

b. It regulates

c. It protects against

B. Physical Characteristics of Blood   (page 377)

B1. Blood is *(thinner? thicker?)* than water and is therefore *(more? less?)* viscous.

B2. Complete the statements below that describe the physical characteristics of blood.

a. Blood has a temperature of about _____ °C, a normal pH range of _____ to _____,

and its salt concentration is _____ percent.

b. Blood constitutes about _____ % of total body weight. Its volume in an average-sized

male is _____ to _____ liters, and _____ to _____ liters in an average-sized
female.

C. Components of Blood (pages 378–379)

C1. Blood is composed of two portions, _____ _____

(cells and cell-like structures) and _____ (liquid containing dissolved

substances). The former composes about _____ % of the volume of blood and the

latter about _____ %.

C2. Complete the table below pertaining to the most common classification of the formed
elements.

a. _____ (red blood cells)
b. _____ (white blood cells)
c. _____ leukocytes (granulocytes)
Neutrophils
d. _____
e. _____
f. Agranular leukocytes   (_____)
Lymphocytes (T cells and B cells)
g. _____
h. Platelets

C3. Test your knowledge of the plasma. (See Exhibit 12.1.)

a. Water constitutes approximately _____ % of the plasma while other solutes, most of

which by weight are proteins, constitute about _____ % of the plasma.

b. List four inorganic cations and anions found in the plasma.

Cations   **Anions**

1. _____   5. _____

2. _____   6. _____

3. _____   7. _____

4. _____   8. _____

c. Name six substances under the heading of wastes.

1. _____   4. _____

2. _____   5. _____

3. _____   6. _____

C4. Match the terms below and the percentages in the plasma with the corresponding description. (See Exhibit 12.1, and the textbook section on blood plasma.) (Fill in one space with the protein abbreviation and the other space with the percentage of that protein found in the plasma.)

AL	albumin	7%
GL	globulin	38%
FB	fibrinogen	54%

a. _____ _____ Produced by the liver, it plays an essential role in blood clotting.

b. _____ _____ Protein group to which antibodies belong.

c. _____ _____ Smallest plasma proteins, produced by the liver. They exert considerable osmotic pressure, which helps maintain water balance between blood and tissue.

D. Formation of Blood Cells   (page 379)

D1. The process by which blood cells are formed is called _____ or

_____.

D2. Contrast the sites of blood cell production in the fetus and the adult.

a. Fetus

b. Adult

D3. The precursor to the five types of cells from which the major types of blood cells develop

is called the _____ stem cell.

E. Erythrocytes (Red Blood Cells)   (pages 380–383)

E1.   Refer to Figure LG 12.1 on p. 140 and answer these questions about erythrocyte structure.

a.   Erythrocytes appear as _____ discs averaging about _____ in

diameter. Which diagram represents a mature erythrocyte? _____

b.   Erythrocytes *(have? lack?)* a nucleus and organelles.

c.   The pigment within the cytoplasm of an RBC, which is responsible for the red color of

whole blood, is called _____. It constitutes approximately _____ % of

the cell weight and functions to carry _____ and

_____ _____.

d.   Label the diagram in Figure LG 12.1 that corresponds to an erythrocyte.

E2.   Test your knowledge of erythrocyte function by answering these questions

a.   Hemoglobin transports about _____ % of the body's total carbon dioxide, a waste
product of metabolism.

b.   A hemoglobin molecule consists of a protein called globin, which is composed of four

_____ chains plus four nonprotein pigments called

_____.

c.   A biconcave disc has a *(greater? lesser?)* surface area for its volume than a cube of the
same volume.

d.   Erythrocytes *(do not? do?)* consume any of the oxygen that they transport.

e.   It has been estimated that each erythrocyte contains about *(200? 280?)* million hemoglobin
molecules.

f.   In the lungs, hemoglobin acquires a fresh supply of oxygen and *(nitric oxide? super nitric
oxide?)*.

E3.   Red blood cells live only about _____ days. A healthy male has about _____ million

red blood cells per cubic millimeter (mm^3) of blood and a female has about _____
million.

E4.   The process by which erythrocytes are formed is called _____.

E5.   Polychromatophilic erythroblasts develop into _____ erythroblasts.

E6.   The immediate precursor to an erythrocyte is a *(proerythroblast? reticulocyte?)*.

E7.   Hypoxia (reduced oxygen delivery) stimulates the _____ to increase

the release of the hormone _____. This hormone *(speeds up? slows
down?)* the development of proerythroblasts into reticulocytes.

E8.   Surface antigens on erythrocytes are called _____. The two major

blood group classifications are the ABO and _____ blood grouping systems.

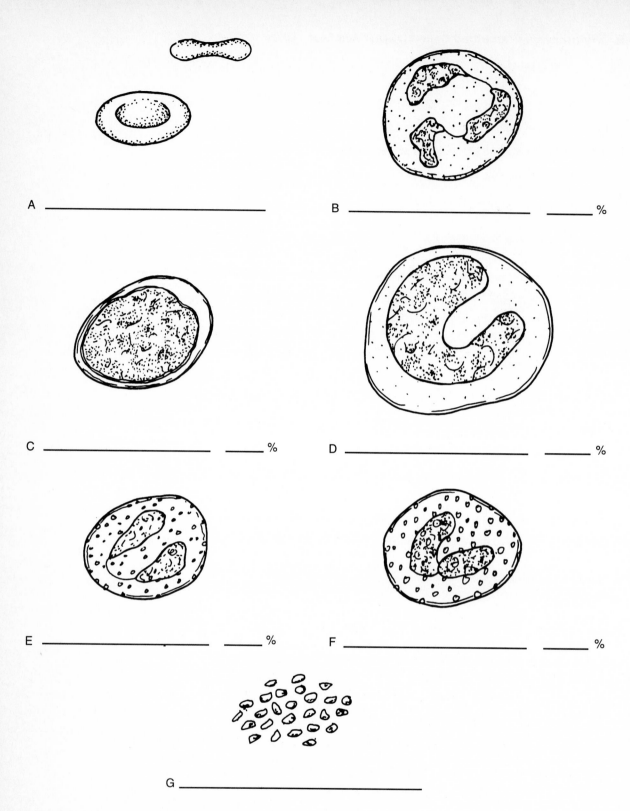

A _____

B _____ _____ %

C _____ _____ %

D _____ _____ %

E _____ _____ %

F _____ _____ %

G _____

Figure LG 12.1 Diagrams of blood cells.

E9. Blood donors who manufacture neither agglutinogen and lack the D antigen would be

classified as blood type _____ _____, while donors who manufacture both

agglutinogens and have the Rh (D) antigen would have blood type _____ _____.

F. Leukocytes (White Blood Cells)   (pages 383–386)

F1. Name two differences between leukocytes and erythrocytes.

a.

b.

F2. Refer to Figure LG 12.1 and the textbook section on leukocytes, and complete this exercise.

a. Label each leukocyte (WBC) and indicate what percentage of the total WBC count is accounted for by each type.

b. Which three cells are granulocytes? _____ _____

_____

c. Which two cells are agranulocytes? _____ _____

F3. Match the leukocyte with the best description (you may use an answer more than once).

B	basophils
E	eosinophils
L	lymphocytes
M	monocytes
N	neutrophils

a. _____ Nuclei are usually kidney shaped or horse-shoe shaped.

b. _____ Have 2–5 lobed nuclei, connected by thin strands.

c. _____ Nuclei are bilobed or irregularly shaped, often in the form of a letter S.

d. _____ Cytoplasm stains sky blue.

e. _____ Bilobed nuclei, connected by thin strand or thick isthmus.

f. _____ Pale lilac colored granules in cytoplasm when stained.

g. _____ Nuclei are darkly stained, round or slightly indented.

h. _____ Cytoplasmic granules stain blue-purple.

i. _____ Cytoplasm is blue-gray with a foamy appearance.

j. _____ Granules stain red-orange.

F4. How do major histocompatibility (MHC) antigens affect the success rate of transplants?

F5. Answer the following questions about leukocyte function.

a. Neutrophils and monocytes are active in _____.

b. A high percentage of monocytes in the blood could be indicative of a(n) *(acute? chronic?)* infection.

c. Monocytes differentiate into phagocytes called _____

_____.

d. A high eosinophil count frequently indicates a(n) *(bacterial? allergic?)* condition.

e. Basophils leave the capillaries, enter the tissues, and develop into mast cells, which liberate

_____, _____, and _____.

F6. Complete the following questions pertaining to antigens and antibodies.

a. A substance that is capable of stimulating an immune response is called a(n)

_____.

b. Antigens and antibodies are usually *(nonprotein? protein?)* substances.

c. The *(B? T?)* lymphocytes differentiate into plasma cells when they react with an antigen.

d. Antigens also stimulate other lymphocytes known as T lymphocytes. One group of T cells,

the _____ T cells, are activated by certain antigens and react by destroy-
ing them directly.

e. The test that determines the percentage of each type of leukocyte in the blood is called a

_____ white blood cell count.

F7. Leukocytes are far less numerous than red blood cells, averaging from _____ to

_____ cells per cubic millimeter (mm³).

F8. An increase in the number of leukocytes is called _____ and a

decrease is called _____.

F9. The ratio of red blood cells to white blood cells is about

a. 600:1   c. 800:1

b. 700:1   d. 900:1

F10. Define the following terms:

a. Chemotaxis

b. Emigration

F11. Contrast the production sites of the granulocytes and agranulocytes.

G. Platelets (page 386)

G1. In Figure LG 12.1, which diagram represents the platelets? _____.

G2. Answer (T) true or (F) false to the questions below.

a. _____ Platelets are whole cells.

b. _____ Platelets' primary function is related to blood clotting.

c. _____ The average life span of a platelet is 3–6 days.

d. _____ A normal platelet count range is 250,000–400,000 per cubic millimeter.

e. _____ Platelets are produced in the lymphoid tissue.

H. Applications to Health (Refer to *Applications to Health* companion to answer Part H questions.)

H1. Define anemia.

H2. Match the anemia type with its description.

AA	aplastic anemia	IA	iron deficiency anemia
HmA	hemolytic anemia	PA	pernicious anemia
HrA	hemorrhagic anemia	ScA	sickle-cell disease

a. _____ Thalassemia represents this type of anemia.

b. _____ The most prevalent anemia.

c. _____ An excessive loss of erythrocytes through bleeding.

d. _____ The destruction or inhibition of the red bone marrow would result in this.

e. _____ This anemia is characterized by the production of an abnormal kind of hemoglobin, which may lead to cell rupture.

f. _____ Results from an inability of the stomach to produce intrinsic factor, which is necessary for vitamin B_{12} absorption.

H3. Differentiate between acute and chronic leukemia.

H4. Infectious mononucleosis occurs mainly in *(younger? older?)* individuals. It is caused by the *(Epstein–Barr? Burkitt–Hodgkin?)* virus. The infection causes the *(T? B?)* lymphocytes to *(shrink? enlarge?)* and become abnormal in appearance.

ANSWERS TO SELECT QUESTIONS

A1. (a) blood; (b) heart; (c) blood vessels.
A2. Interstitial fluid.
B1. Thicker, more.
B2. (a) 38, 7.35, 7.45, .90; (b) 8, 5, 6, 4, 5.
C1. formed elements, plasma, 45%, 55%.
C2.

a.	**Erythrocytes**	(red blood cells)
b.	**Leukocytes**	(white blood cells)
c.	**Granular**	(granulocytes)
	Neutrophils	
d.	**Eosinophils**	
e.	**Basophils**	
f.	Agranular leukocytes	**(agranulocytes)**
	Lymphocytes (T cells and B cells)	
g.	**monocytes**	
h.	Platelets	

C3. (a) 91.5, 8.5; (b) 1—Na^+, 2—K^+, 3—Ca^{2+}, 4—Mg^{2+}, 5—Cl^-, 6—HPO_4^{2-}, 7—SO_4^{2-}, 8—HCO_3^-; (c) urea, uric acid, creatine, creatinine, bilirubin, ammonium salts.
C4. (a) FB, 7%; (b) GL, 38%; (c) AL, 54%.
D1. Hematopoiesis, hemopoiesis.
D2. (a) Yolk sac, liver, spleen, thymus gland, lymph nodes, bone marrow; (b) red blood marrow in the proximal epiphyses of the humerus and femur, flat bones (cranium and sternum), vertebrae, pelvis.
D3. Pluripotential.

E1. (a) Biconcave, 7–8 μm, A; (b) lack; (c) hemoglobin, 33, oxygen, carbon dioxide.
E2. (a) 23; (b) polypeptide, heme; (c) greater; (d) do not; (e) 280; (f) super nitric oxide.
E3. 120, 5.4, 4.8.
E4. Erythropoiesis.
E5. Orthochromatophilic.
E6. Reticulocyte.
E7. Kidney, erythropoietin, speeds up.
E8. Agglutinogens, Rh.
E9. O (Rh^-), AB (Rh^+).
F1. (a) Leukocytes have a nucleus, erythrocytes do not; (b) erythrocytes possess hemoglobin, leukocytes do not.
F2. (b) Basophils, eosinophils, neutrophils; (c) lymphocytes, monocytes.
F3. (a) M; (b) N; (c) B; (d) L; (e) E; (f) N; (g) L; (h) B; (i) M; (j) E.
F5. (a) phagocytosis; (b) chronic; (c) wandering macrophages; (d) allergic; (e) heparin, histamine, serotonin.
F6. (a) Antigen; (b) protein; (c) B; (d) (cytotoxic) killer; (e) differential.
F7. 5,000, 10,000.
F8. Leukocytosis, leukopenia.
F9. b.
G1. G.
G2. (a) F; (b) T; (c) F; (d) T; (e) F.
H2. (a) HmA; (b) IA; (c) HrA; (d) AA; (e) ScA; (f) PA.
H4. Younger, Epstein–Barr, B, enlarge.

SELF QUIZ

Choose the one best answer to the following questions.

1. A reticulocyte is a cell involved in the formation of _____.

   A. red blood cells   C. neutrophils
   B. basophils   D. platelets

2. This fraction of plasma proteins accounts for 54% of the total amount: _____.

   A. globulin   C. albumin
   B. fibrinogen   D. prothrombin

3. Megakaryoblasts form into mature

   _____.

   A. erythrocytes   C. platelets
   B. neutrophils   D. monocytes

4. Red blood cells live only about _____ days.

   A. 110   C. 130
   B. 120   D. 140

5. This white blood cell functions by combating the effects of histamines in allergic reactions.

   A. neutrophil   D. monocyte
   B. basophil   E. lymphocyte
   C. eosinophil

6. Which cells are responsible for the secretion of antibodies?

   A. basophils   D. B lymphocytes
   B. eosinophils   (plasma cells)
   C. T lymphocytes

7. Which type of anemia is associated with an inability to produce intrinsic factor?

   A. iron deficiency
   B. pernicious anemia
   C. hemorrhagic anemia
   D. hemolytic anemia
   E. aplastic anemia

8. Which of the following statements is NOT true?

   A. Blood is more viscous than water.
   B. Blood makes up approximately 10% of body weight.
   C. The pH of blood is 7.35–7.45.
   D. In an average-sized male there are about 5–6 liters of blood.

9. The normal RBC count in a healthy male is _____ RBCs/mm^3.

   A. 4.8 million
   B. 250,00–400,000
   C. 5,000–10,000
   D. 5.4 million

10. An individual whose red blood cells lack both agglutinogens would possess which blood type?

   A. A
   B. B
   C. AB
   D. O

11. Which of the following cells does NOT belong with the others?

   A. basophil
   B. eosinophil
   C. monocyte
   D. neutrophil

12. This disorder is characterized by excess deposit of iron in the tissues, especially the liver and pancreas.

   A. aplastic anemia
   B. infectious mononucleosis
   C. hemochromatosis
   D. sickle cell disease

13. This white blood cell has a normal differential value of between 20 and 25%.

   A. neutrophils
   B. lymphocytes
   C. monocytes
   D. basophils

14. Which of the following values represent the white blood cells per cubic millimeter associated with leukopenia?

   A. 11,000
   B. 6,000
   C. 4,000
   D. 13,000

15. The precursor (stem cell) of all blood cells is the _____.

   A. myeloblast
   B. pluripotential
   C. promegakaryocyte
   D. proerythroblast
   E. reticulocyte

Fill in the blanks.

16. The kidneys release _____ in response to reduced oxygen delivery to tissues (hypoxia).

17. Hemoglobin transports about _____ percent of the body's total carbon dioxide.

18. A very low platelet count, which results in a tendency to bleed from the capillaries, is referred to as _____.

19. The primary function of a platelet is to initiate a chain of reactions that result in _____ _____.

20. _____ refers to a decreased delivery of oxygen to the tissues.

ANSWERS TO THE SELF QUIZ

1. A
2. C
3. C
4. B
5. C
6. D
7. B
8. B

9. D
10. D
11. C
12. C
13. B
14. C
15. B
16. Erythropoietin

17. 23
18. Thrombocytopenia
19. Blood clotting
20. Hypoxia

The Cardiovascular System: The Heart

CHAPTER
13

SYNOPSIS

It starts beating approximately 22 days after conception. It contracts over 100,000 times per day without getting tired and can do this for 100 years or more. As the center of the cardiovascular system, this hollow muscular organ, which is only slightly bigger than your fist, has the job of pumping 14,000 liters (3,600 gallons) of blood through more than 100,000 kilometers (60,000 miles) of blood vessels.

Unlike the hearts of many lower animals, our heart is designed in a manner that allows for the total separation of **oxygenated** and **deoxygenated** blood. This separation provides our body with the most effective method of supplying oxygen to, and removing carbon dioxide from, our cells. This design surely makes the heart one of the most efficient and expertly conceived organs of the body.

Your journey now takes you inside one of the most marvelous structures in nature: the **human heart.**

TOPIC OUTLINE AND OBJECTIVES

A. Location and Surface Projection

1. Discuss the size of the heart and its location relative to the internal and surface anatomy.

B. Pericardium

2. List and describe the location and function of the components that form the coverings of the heart.

C. Heart Wall

3. List and describe the location and function of the three layers of the heart wall: epicardium, myocardium, and endocardium.

D. Chambers of the Heart

4. Compare the location and features of the four chambers located within the heart.
5. Identify the major vessels that attach to the heart.

E. Valves of the Heart and Heart Sounds

6. Identify, locate, and describe the function and features of the four valves of the heart.
7. Describe the skeleton and surface projections of the heart.

F. Heart Blood Supply

8. Discuss the route of blood in the coronary circulation.

G. Conduction System and Pacemaker

9. Explain the structural and functional features of the conduction system.

H. Cardiac Cycle (Heartbeat)

I. Exercise and the Heart

J. Developmental Anatomy of the Heart

K. Applications to Health

10. Describe how atherosclerosis and coronary artery spasm contribute to coronary artery disease (CAD).

11. Define atherosclerosis, coronary artery spasm, patent ductus arteriosus, septal defects, valvular stenosis, tetralogy of Fallot, arrhythmias, heart block, flutter, and fibrillation.

N. Key Medical Terms Associated with the Heart

SCIENTIFIC TERMINOLOGY

Find an anatomical sample word for each prefix and suffix:

Prefix/Suffix	Meaning	Sample Word
angio-	vessel	
cardio-	heart	
chorda-	cord	
cor-	heart	
endo-	within	
epi-	on top	
mega-	large	
myo-	muscle	
peri-	around	

A. Location and Surface Projection of the Heart   (page 393–394)

A1. The heart rests on the diaphragm near the midline of the thoracic cavity, in a mass of

tissue called the _____. Approximately _____ of the heart's mass lies to the *(left? right?)* of the body's midline.

A2. The pointed end of the heart is called the *(base? apex?),* while the wide superior and

posterior margin is called the _____.

A3. The *(diaphragmatic? sternocostal?)* surface is deep to the sternum and ribs

A4. The *(superior? inferior?)* *(left? right?)* point is located at the inferior border of the second left costal cartilage.

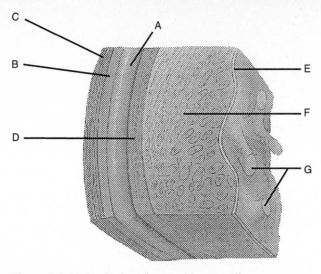

Figure LG 13.1   Pericardium and heart wall.

B. Pericardium   (pages 394–396)

B1. Refer to Figure LG 13.1 and match the letters with the following components of the pericardium.

_____ fibrous pericardium

_____ parietal layer (serous pericardium)

_____ visceral layer (serous pericardium)

_____ pericardial cavity

B2. The thin film of serous fluid located in the pericardial cavity is known as the

_____ fluid. Its function is to *(increase? decrease?)* friction between the membranes as the heart beats.

B3. The pericardium is a *(double? triple?)*-layered bag that surrounds and protects the heart.

C. Heart Wall   (page 396)

C1. Refer to Figure LG 13.1 and match the letters with the following structures.

_____ epicardium   _____ myocardium

_____ endocardium   _____ trabeculae carneae

C2. The bulk of the heart wall is formed by the *(epicardium? myocardium? endocardium?)*.

C3. The endocardium is composed of a thin layer of _____ overlying a thin

layer of _____ tissue.

C4. Cardiac muscle fibers are in physical contact with neighboring fibers via transverse

thickenings of the sarcolemma called _____ discs. Within the discs are

_____ junctions that allow the action potential to spread from one fiber
to another.

D. Chambers of the Heart (pages 397–403)

D1. The interior of the heart is divided into _____ chambers. The superior chambers are

called _____ and the inferior chambers are called

_____.

D2. A prominent feature of the interatrial septum is an oval depression called the

_____ _____, which corresponds to the former site

of the _____ _____.

D3. The _____ _____ are an irregular surface of ridges
and folds of the myocardium, covered by endocardium, which are located in the
ventricles.

D4. Refer to Figure LG 13.2 to do the following activities.

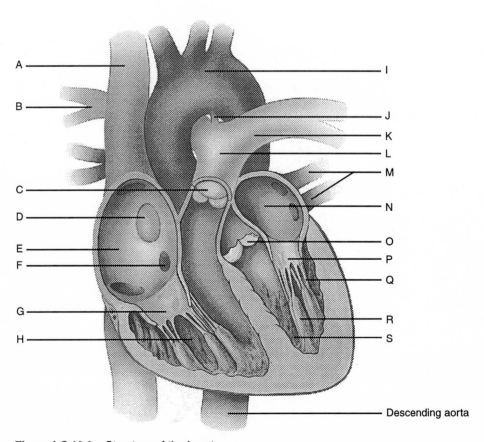

Figure LG 13.2 Structure of the heart.

a. Identify all the lettered structures.

A _____   J _____

B _____   K _____

C _____   L _____

D _____   M _____

E _____   N _____

F _____   O _____

G _____   P _____

H _____   Q _____

I _____   R _____

b. Shade red the chambers that contain highly oxygenated blood; shade blue the chambers with low oxygenated blood.

c. Draw arrows to indicate the direction of blood flow through the heart.

D5. The thickest chamber of the heart is the *(right? left?) (atrium? ventricle?)*. Why?

D6. List the three veins that return deoxygenated blood to the heart and describe the region(s) they drain.

a.

b.

c.

D7. Check your understanding of the heart structure and the pathway of blood by completing this matching section.

A	atria	PT	pulmonary trunk
AA	ascending aorta	PV	pulmonary veins
DA	ductus arteriosum	V	ventricles
PA	pulmonary arteries		

a. _____ Returns oxygenated blood from the lungs to the heart.

b. _____ Chambers of heart that receive blood from the body or the lungs.

c. _____ Chambers that pump blood away from the heart.

d. _____ Delivers blood into the pulmonary arteries.

e. _____ Receives blood after ejection from the left ventricle.

f. _____ Carries deoxygenated blood to lungs (two answers).

g. _____ Normally closes shortly after birth.

E. Valves of the Heart and Heart Sounds (pages 403–406)

E1. Test your knowledge of the heart valves by answering the questions below. Note: some answers can be used more than once.

ASL	aortic semilunar valve
BV	bicuspid valve
CT	chordae tendineae
PSL	pulmonary semilunar valve
TV	tricuspid valve

a. _____ Also known as the atrioventricular (AV) valves.

b. _____ Connect the pointed ends and undersurfaces of the AV valves to the papillary muscles.

c. _____ Consist of three crescent-shaped cusps.

d. _____ Also referred to as the mitral valve.

e. _____ Found between the right atrium and right ventricle.

f. _____ Prevents the backflow of blood into the left ventricle.

g. _____ Opens to allow the ejection of blood from the right ventricle.

h. _____ Prevents backflow of blood into the left atrium.

i. _____ Prevents backflow of blood into the right atrium.

E2. Cardiac muscle tissue along with _____ _____ tissue forms the skeleton of the heart.

E3. What is the skeleton of the heart? List the four components associated with it.

a.

b.

c.

d.

E4. Match the valve location with its surface projection.

> A aortic semilunar valve
> B bicuspid valve
> P pulmonary semilunar valve
> T tricuspid valve

a. _____ Lies horizontally posterior to the medial end of the left third costal cartilage and the adjoining part of the sternum.

b. _____ Lies posterior to the left side of the sternum obliquely at the level of the fourth costal cartilage.

c. _____ Lies posterior to the sternum, extending from the midline at the level of the fourth costal cartilage down toward the right sixth chondrosternal junction.

d. _____ Is placed obliquely posterior to the left side of the sternum at the level of the third intercostal space.

E5. _____ is the act of listening to sounds within the body.

E6. The sounds produced during a cardiac cycle occur from the *(opening? closing?)* of the valves.

E7. The first heart sound, described as _____, is created by the *(AV? semilunar?)* valves.

E8. Each cardiac cycle generates *(2? 4?)* heart sounds.

F. Heart Blood Supply (page 406)

F1. Why does the heart need its own blood supply when it has all the oxygen and nutrients it needs passing through its chambers?

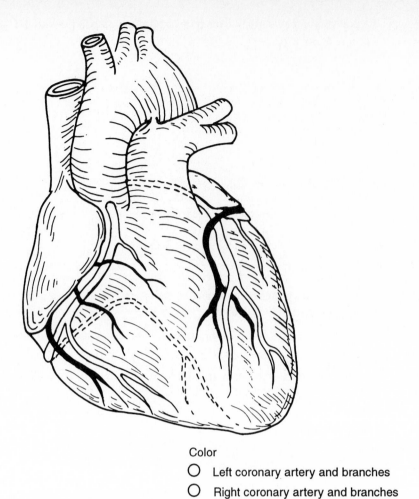

Color

○ Left coronary artery and branches
○ Right coronary artery and branches

Figure LG 13.3 Anterior view of the heart.

F2. Refer to Figure LG 13.3 and answer the following questions.

a. Color the left coronary artery and its branches. Color the right coronary artery and its branches.

b. The left coronary artery divides into the

1.

2.

c. The right coronary artery divides into the

1.

2.

F3. The principal tributaries carrying blood into the coronary sinus are the

_____ _____ vein and the _____

_____ cardiac vein.

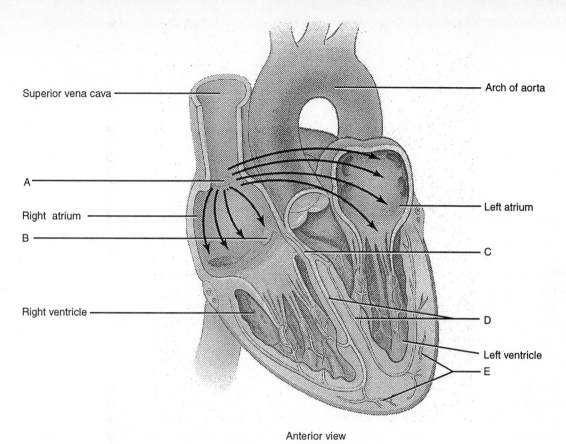

Superior vena cava

Arch of aorta

A

Right atrium

B

Left atrium

C

Right ventricle

D

Left ventricle

E

Anterior view

Figure LG 13.4   Conduction systems of the heart.

F4. What is an anastomosis? Why is it so important to the heart?

G. Conduction System and Pacemaker   (pages 406–409)

G1. Refer to Figure LG 13.4 and match the letters with the conduction system structure, listed below.

_____ atrioventricular bundle

_____ atrioventricular node

_____ conduction myofibers

_____ right and left bundle branches

_____ sinoatrial node

G2. Each cardiac muscle fiber is in physical contact with other fibers by transverse

thickenings of the sarcolemma called _____ discs.

G3. Within these discs are gap _____ that aid in the conduction of muscle action potentials between cardiac muscle fibers.

G4. The autonomic nervous system *(does? does not?)* initiate cardiac muscle contraction.

G5. The compact mass of cells located in the right atrial wall inferior to the opening of the

superior vena cava is called the _____ _____ or
pacemaker.

G6. The autonomic nervous system *(can? cannot?)* influence the depolarization of the
pacemaker.

G7. The _____ _____ is located in the interatrial septum
and is one of the last portions of the atria to be depolarized, which allows time for the
atria to empty their blood into the ventricles before the ventricles begin their contraction.

G8. Actual contraction of the ventricles is stimulated by the _____

_____ .

H. Cardiac Cycle (Heartbeat)   (pages 409–410)

H1. Match the following cardiac cycle phase with the proper description.

RP	relaxation period
VF	ventricular filling
VS	ventricular systole

a.   _____ This phase is represented by the QRS wave in the EKG.

b.   _____ Throughout this phase the AV valves are open and the semilunar valves are
closed.

c.   _____ At the beginning of this phase, all four chambers are in diastole.

I. Exercise and the Heart   (page 410)

I1. In addition to increasing cardiac output, what other benefits of physical conditioning
occur with exercise?

J. Developmental Anatomy of the Heart   (page 410)

J1. The heart derives from the _____ before the end of the *(third? fourth?)*
week.

J2. List the five regions of the primitive heart tube.

a.

b.

c.

d.

e.

K. Applications to Health (Refer to *Application to Health* companion to answer Part K questions.)

K1. Answer the following questions pertaining to atherosclerosis.

a. The first event in atherosclerosis is thought to be damage of the _____

_____ of the artery.

b. List two events that follow the damage that occurred in question (a) above.

　1.

　2.

c. The accumulation of cholesterol, fatty substances, triglycerides, and smooth muscle fibers

form a mass called an _____ plaque.

K2. Match the following treatments with their description.

CABG	coronary artery bypass grafting
PTCA	percutaneous transluminal coronary angioplasty
ST	stent

a. _____ A balloon catheter is inserted into an artery and guided through the arterial system. The balloon is inflated, compressing the obstruction against the arterial wall.

b. _____ A portion of a blood vessel is removed from another part of the body and it is grafted between the aorta and the unblocked portion of the coronary artery.

c. _____ A special device is inserted via a catheter to keep an artery patent (open).

K3. Match the congenital defect with its description.

IVD	interventricular septal defect
PDA	patent ductus arteriosus
TOF	tetralogy of Fallot
VS	valvular stenosis

a. _____ A narrowing of one of the valves regulating blood flow.

b. _____ Caused by incomplete closure of the interventricular septum.

c. _____ A temporary blood vessel between the aorta and the pulmonary trunk does not close after birth.

d. _____ A combination of four defects: an interventricular septal defect, an aorta that emerges from both ventricles, a stenosed pulmonary semilunar valve, and an enlarged right ventricle.

K4. _____ refers to an abnormality or irregularity in the heart rhythm. A

very serious type is called _____ fibrillation (VF). It almost always indicates imminent death unless corrected.

K5. The most common heart blockage occurs in the _____

_____.

K6. Atrial (*flutter? fibrillation?*) is described as an uncoordinated contraction of the atrial muscles that cause the atria to contract irregularly and still faster.

ANSWERS TO SELECT QUESTIONS

A1. Mediastinum, two-thirds, left.
A2. Apex, base.
A3. Sternocostal.
A4. Superior, left.
B1. A—Pericardial cavity, B—parietal layer,
C—fibrous pericardium, D—visceral layer.
B2. Pericardial, decrease.
B3. Triple.
C1. D—epicardium, E—endocardium,
F—myocardium, G—trabeculae carneae.
C2. Myocardium.
C3. Endothelium, connective.
C4. Intercalated, gap.
D1. 4, atria, ventricles.
D2. Fossa ovalis, foramen ovale.
D3. Trabeculae carneae.
D4. (a) A—Superior vena cava, B—right pulmonary
artery, C—pulmonary semilunar valve, D—fossa
ovalis, E—right atrium, F—opening of the coro-
nary sinus, G—tricuspid valve, H—right ventri-
cle, I—arch of the aorta, J—ligamentum
arteriosum, K—left pulmonary artery,
L—pulmonary trunk, M—left pulmonary veins,
N—left atrium, O—aortic semilunar valve,
P—bicuspid valve, Q—chordae tendineae,
R—papillary muscle, S—left ventricle.
D5. Left, ventricle; pumps blood to all body regions
except the lungs.
D6. (a) Superior vena cava—drains areas superior to
heart; (b) inferior vena cava—drains areas inferior
to diaphragm; (c) coronary sinus—drains blood
from heart wall.
D7. (a) PV; (b) A; (c) V; (d) PT; (e) AA; (f) PA and PT;
(g) DA.

E1. (a) TV, BV; (b) CT; (c) PSL, ASL; (d) BV; (e) TV;
(f) ASL; (g) PSL; (h) BV; (i) TV.
E2. Dense connective.
E4. (a) P; (b) B; (c) T; (d) A.
E5. Auscultation.
E6. Closing.
E7. Lubb, AV.
E8. 4.
F2. (b) 1—Anterior interventricular or anterior
descending arteries, 2—circumflex branch;
(c) 1—posterior interventricular branch,
2—marginal branch.
F3. Great cardiac, middle.
G1. A—Sinoatrial node; B—atrioventricular node;
C—atrioventricular bundle; D—right/left bundle
branches; E—conduction myofibers.
G2. Intercalated.
G3. Junctions.
G4. Does not.
G5. Sinoatrial node.
G6. Can.
G7. Atrioventricular node.
G8. Conduction myofibers.
H1. (a) VS; (b) VF; (c) RP.
J1. Mesoderm, third.
J2. (a) truncus arteriosus; (b) bulbus cordis; (c) ventri-
cle, (d) atrium, (e) sinus venosus.
K1. (a) endothelial lining; (b) 1—smooth muscle fiber
proliferation, 2—lipid builds-up, both within cells
and the interstitial spaces; (c) atherosclerotic.
K2. (a) PTCA; (b) CABG; (c) ST.
K3. (a) VS; (b) IVD; (c) PDA; (d) TOF.
K4. Arrhythmia, ventricular.
K5. Atrioventricular (AV) node.
M6. Fibrillation.

SELF QUIZ

Choose the one best answer to the following questions.

1. The thickest layer of the heart wall is the

   A. pericardium   D. myocardium
   B. epicardium   E. none of the
   C. endocardium   above are correct

2. The space between the parietal and visceral layers
   of the serous pericardium is filled with

   A. fat   D. serous fluid
   B. mucous   E. cartilage
   C. water

3. Blood ejected from the right ventricle passes
   through the _____ valve.

   A. mitral   C. tricuspid
   B. pulmonary semi-   D. bicuspid
      lunar   E. aortic semilunar

4. The chordae tendineae attach from the

   A. wall of the atria to the atrioventricular valves
   B. interventricular septum to the semilunar valves
   C. left atrium to the right atrium via the foramen
      ovale
   D. tips of the atrioventricular valves to the papil-
      lary muscles
   E. atrioventricular valves to the semilunar valves

5. The foramen ovale directs blood, in the fetal heart, from the

   A. left atrium to the right atrium
   B. right ventricle to the left ventricle
   C. right ventricle to the right atrium
   D. right atrium to the left atrium
   E. pulmonary trunk to the aorta

6. Which valve closes when the left ventricle is relaxing?

   A. aortic semilunar
   B. mitral
   C. tricuspid
   D. bicuspid
   E. B and C are both correct answers

7. Which of the following is correctly paired?

   A. fibrous rings—formed by the fusion of the fibrous connective tissue of the pulmonary and aortic rings
   B. left fibrous trigone—supports the four valves of the heart
   C. right fibrous trigone—large triangular mass formed by left atrioventricular, aortic, and right atrioventricular rings
   D. conus tendon—small mass formed by atrioventricular and aortic fibrous rings
   E. none of the above answers are correct

8. The right coronary artery divides into which branches?

   A. anterior interventricular
   B. circumflex
   C. posterior interventricular
   D. marginal
   E. A and B
   F. C and D

9. The sinoatrial node is

   A. the pacemaker of the heart
   B. requires nervous innervation in order to contract
   C. is located in the inferior wall of the right atrium
   D. A and C are correct
   E. none of the above are correct

10. The first heart sound is produced by the

    A. closure of the atrioventricular valves
    B. closure of the semilunar valves
    C. opening of the atrioventricular valves
    D. opening of the semilunar valves
    E. A and C are correct

11. The backflow of blood from the left ventricle to the left atrium could be caused by

    A. aortic stenosis
    B. mitral valve insufficiency
    C. aortic insufficiency
    D. mitral stenosis
    E. tricuspid valve prolapse

12. Chest pain resulting from ischemia of the myocardium is called _____.

    A. myocardial infarction
    B. angina pectoris
    C. myocarditis
    D. cardiac tamponade
    E. heart block

13. An irregular surface of ridges and folds of the myocardium covered by endocardium in the ventricles is known as the

    A. papillary muscles
    B. pectinate muscles
    C. intercardial muscles
    D. pectina papillae
    E. trabeculae carneae

14. The thin layer of endothelium that lines the inside of the heart is called the

    A. pericardium
    B. endocardium
    C. epicardium
    D. myocardium
    E. B and C are both correct

15. The coronary sulcus

    A. is a groove that separates the right and left ventricles
    B. contains a coronary blood vessel
    C. contains a variable amount of fat
    D. separates the atria from the ventricles
    E. C and D are both correct

Answer (T) True or (F) False to the following questions.

16. _____ A narrowing of a valve is called a stenosis.

17. _____ The bicuspid (mitral) valve is located between the right atrium and right ventricle.

18. _____ In atrial flutter the atrial rhythm averages about 300 beats per minute.

19. _____ Opening of the semilunar valves is responsible for the dupp portion of the heart sound.

20. _____ Blood ejected from the left ventricle enters the pulmonary trunk.

Fill in the blanks.

21. Impulses from the sinoatrial node travel next to the _____ _____.

22. Physiological _____ would refer to the enlargement of a long-distance runner's heart.

23. A period of rapid heartbeats that begin and end suddenly is called _____

_____.

24. _____ is a fluttering of the heart or abnormal rate or rhythm of the heart.

25. _____ _____ refers to right ventricular hypertrophy from disorders that bring about hypertension in the pulmonary circulation.

ANSWERS TO THE SELF QUIZ

1. D
2. D
3. B
4. D
5. D
6. A
7. C
8. F
9. A

10. A
11. B
12. B
13. E
14. B
15. E
16. T
17. F
18. T

19. F
20. F
21. Atrioventricular node
22. Cardiomegaly
23. Paroxysmal tachycardia
24. Palpitation
25. Cor pulmonale

The Cardiovascular System: Blood Vessels

SYNOPSIS

With the heart as the center of the cardiovascular system, it is the blood vessels that become the "highway" that allows for the distribution of blood throughout the body. In much the same way that the highways we drive on branch into secondary and tertiary roadways, so do the blood vessels begin as large vessels, eventually branching into microscopic vascular networks.

Traversing nearly 100,000 kilometers (60,000 miles), blood vessels are found near almost every cell in the body. **Arteries** are vessels that transport blood from the heart to the tissues or the lungs. Beginning as large vessels, the arteries divide into medium-sized arteries and then into smaller arteries called **arterioles.** As arterioles enter a tissue, they branch into a myriad of microscopic vessels called **capillaries.** It is at the capillary level that the diffusion of gases, nutrients, and wastes occurs. Before exiting the tissues, capillaries reunite to form small veins called **venules,** which combine into larger vessels called **veins.** The veins conduct the blood back to the heart.

As your journey through the body continues, you will take a drive on a true "superhighway." Marvel as you leave the heart, traveling out the aorta, through miles of blood vessels into microscopic capillary beds, some so small that the red blood cells, which are about 8 microns in diameter, must pass through in single file.

Your return trip takes you through the veins that will return to your starting point, the heart.

TOPIC OUTLINE AND OBJECTIVES

A. Arteries, Arterioles, Capillaries, Venules, and Veins

1. Compare elastic (conducting) and muscular (distributing) arteries.
2. Define anastomoses.
3. Contrast the structure and function of arterioles, capillaries, venules, and veins.

B. Blood Distribution

4. Define a blood reservoir and discuss the distribution of blood in the body.

C. Circulatory Routes

5. Differentiate between systemic and pulmonary circulation.
6. Discuss hepatic portal circulation and explain its importance.
7. Compare fetal and adult circulation.

D. Developmental Anatomy of Blood Vessels and Blood

E. Aging and the Cardiovascular System

8. Describe the effects of aging on the cardiovascular system.

F. Applications to Health

9. Discuss hypertension, aneurysm, coronary artery disease (CAD), and deep-venous thrombosis (DVT).

G. Key Medical Terms Associated with Blood Vessels

SCIENTIFIC TERMINOLOGY

Find an anatomical sample word for each prefix and suffix:

Prefix/Suffix	Meaning	Sample Word
aer-	air	
hepato-	liver	
hypo-	below	
-itis	inflammation	
ortho-	straight	
phlebo-	vein	
pulmo-	lung	
-tension	pressure	
-thrombo-	clot	

A. Arteries, Arterioles, Capillaries, Venules, and Veins (pages 417–422)

A1. Refer to Figure LG 14.1 and label the following structures.

_____ endothelium

_____ basement membrane

_____ internal elastic lamina

_____ smooth muscle

_____ external elastic lamina

_____ tunica externa

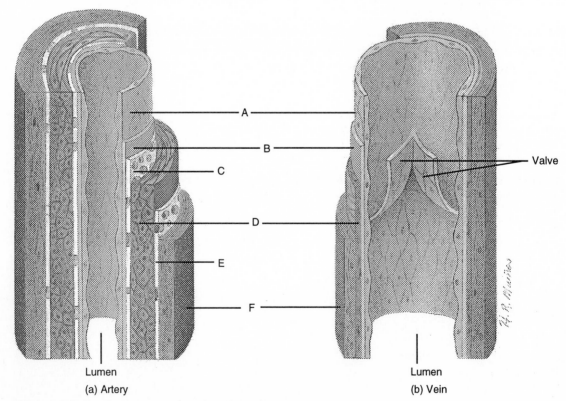

Figure LG 14.1 Comparative structure of blood vessels.

A2. The hollow center of a blood vessel is called the _____.

A3. The middle coat of an artery is called the tunica _____, which consists

of _____ fibers and _____ muscle fibers. This
combination of fibers give the arteries two important functional properties:

_____ and _____.

A4. Sympathetic stimulation to the muscle in most arteries leads to _____,
which is a(n) *(increase? decrease?)* in the size of the lumen.

A5. The outer coat of an arterial wall is composed principally of _____ and

_____ fibers.

A6. List six arteries classified as elastic (conducting) arteries.

a. _____   d. _____

b. _____   e. _____

c. _____   f. _____

A7. The elastic recoil of a large artery creates *(forward? backward?)* flow of blood.

A8. List nine arteries classified as muscular (distributing) arteries.

a. _____   f. _____

b. _____   g. _____

c. _____   h. _____

d. _____   i. _____

e. _____

A9. Define the following terms.
a. Anastomosis

b. Collateral circulation

c. End arteries

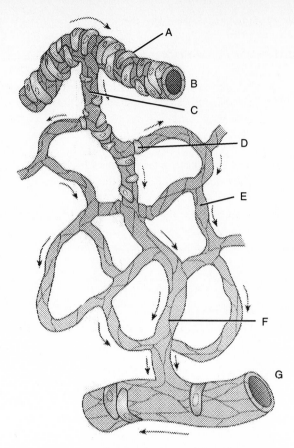

Details of a capillary network

Figure LG 14.2   Arteriole, capillaries, and venule.

A10. Refer to Figure LG 14.2 and label the following structures.

_____ arteriole   _____ thoroughfare channel

_____ metarteriole   _____ true capillary

_____ muscle fiber   _____ venule

_____ precapillary sphincter

A11. Contrast the following.

a. The function of capillaries and thoroughfare channels

b. Continuous capillaries, fenestrated capillaries, and sinusoids

A12. The tunica media of veins is much *(thinner? thicker?)* than that of accompanying arteries.

A13. The blood pressure in veins is *(lower? higher?)* than that in arteries.

A14. The structures located in veins, especially those of the limbs, that prevent the backflow

of blood are called _____.

A15. The _____ _____ cell is an example of a specialized
phagocytic cell that lines a sinusoid.

A16. Two examples of vascular (venous) sinuses are

a.

b.

B. Blood Distribution   (page 422)

B1. Match the percentage of blood found in the following blood vessels.

60%	15%	5%

a. _____ systemic capillaries

b. _____ systemic veins and venules (blood reservoirs)

c. _____ systemic arteries and arterioles

B2. The veins of the abdominal organs, especially the _____ and

_____, are among the principal blood reservoirs.

B3. How do blood reservoirs function? Why is vasoconstriction important?

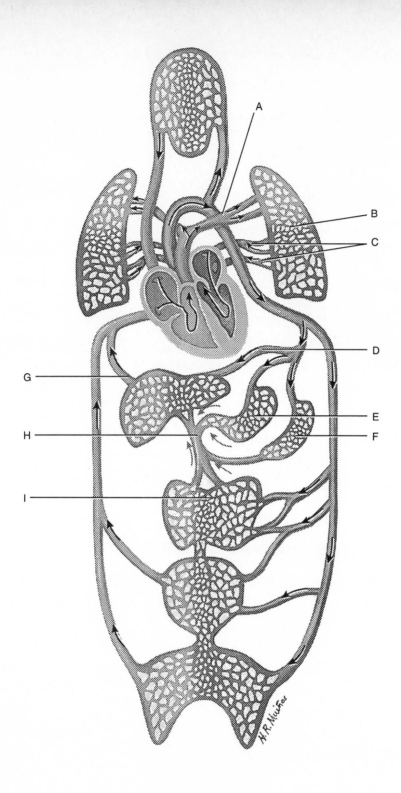

Figure LG 14.3 Circulatory routes.

C. Circulatory Routes (pages 423–471)

C1. Refer to Figure LG 14.3 and answer the following questions.

a. Which letters represent the pulmonary arteries (_____), pulmonary capillaries

(_____), and pulmonary veins (_____)?

b. The letter that represents the capillaries of the spleen is (_____), while (_____)
represents the capillaries of the stomach.

c. Which letter represents the hepatic portal vein (_____)? Which represents the hepatic

vein (_____) and common hepatic artery (_____)?

d. Letter I represents what structure? _____

C2. The _____ _____ includes all the blood vessels that
carry oxygenated blood to the body tissues and return deoxygenated blood to the right
atrium.

C3. Match the following circulatory routes with their description.

> CRC   coronary circulation
> CC   cerebral circulation
> HPC   hepatic portal circulation
> PC   pulmonary circulation

a. _____ Blood flows from the right ventricle to the lungs, and returns to the left atrium.

b. _____ Supplies the brain.

c. _____ Transports blood from the gastrointestinal tract to the liver.

d. _____ Supplies the heart.

Right side
of body

Left side
of body

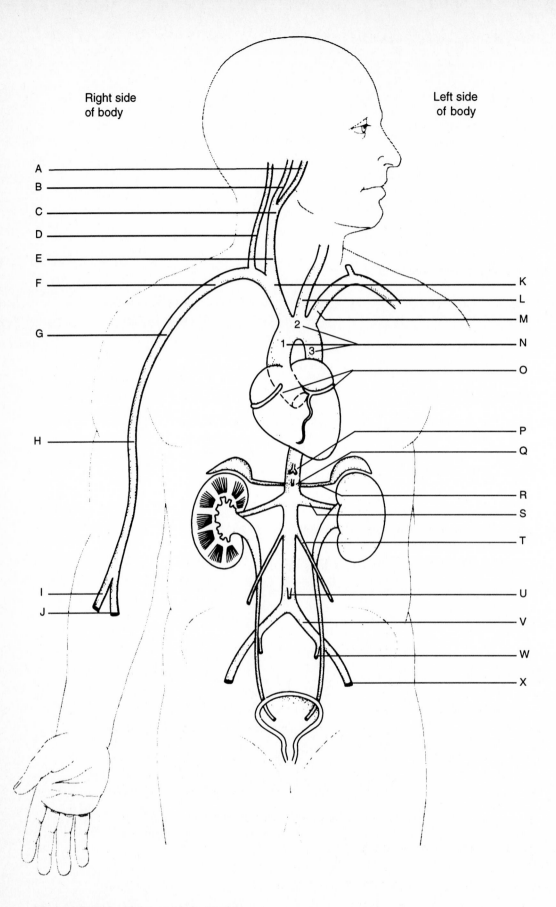

A
B
C
D
E
F

G

H

I
J

K
L
M

2
N
1
3
O

P
Q

R
S

T

U

V

W

X

Figure LG 14.4 Major systemic arteries.

C4. Refer to Exhibit 14.1 in your text and to Figure LG 14.4; then answer the following questions pertaining to the aorta and its branches.

a. The _____ is the largest artery in the body. The first vessels that branch

off this artery are the right and left _____ arteries, which supply blood to

the _____. Which letter corresponds to these vessels? _____

b. Three arteries branch off the arch of the aorta. They are the _____ (letter

_____), left _____ _____ (letter _____), and the

left _____ (letter _____).

c. The _____, superior _____, bronchial, and esophageal
arteries all branch off the thoracic aorta.

d. The _____ artery (trunk) gives rise to the common hepatic, left gastric,
and splenic arteries.

e. The _____ artery supplies the kidneys, while the

_____ mesenteric artery supplies the small intestine, cecum, ascending
and transverse colons, and pancreas.

C5. Refer to Exhibit 14.3 in your textbook and Figure LG 14.4, as you test your knowledge of
the arch of the aorta.

a. The brachiocephalic trunk is the _____ and _____

branch off the arch of the aorta. It divides into the *(right? left?)* subclavian (letter _____)

artery and common carotid artery (letter _____).

b. The subclavian artery, upon reaching the shoulder, is called the _____

artery (letter _____), which becomes the _____ artery (letter

_____) in the arm.

c. The ulnar (letter _____) and radial (letter _____) arteries anastomose to form two

palmar arches called the _____ _____ and the

_____ _____.

d. The major branch off the subclavian artery which passes through the transverse foramina of

the cervical vertebrae is the _____ artery (letter _____). This artery

unites with its counterpart, from the other side, to form the _____ artery.

e. The *(internal? external?)* carotid arteries supply the throat, face, and scalp (letter _____).

f. The *(internal? external?)* carotid arteries (letter _____) unite with the basilar artery to

form the _____ _____ _____.

g. The carotid bifurcation is indicated by letter _____.

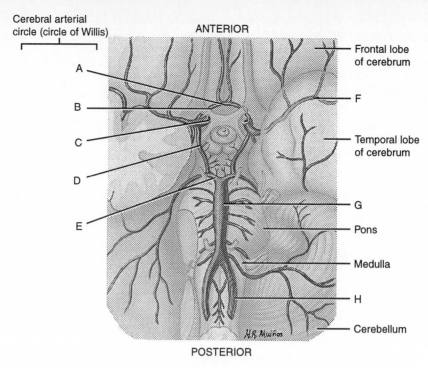

Cerebral arterial circle (circle of Willis)

ANTERIOR

A

B

C

D

E

Frontal lobe of cerebrum

F

Temporal lobe of cerebrum

G

Pons

Medulla

H

Cerebellum

POSTERIOR

Inferior view of base of brain

Figure LG 14.5   Principal arteries of the base of the brain.

C6.   Refer to Figure LG 14.5, labeling the following arteries.

_____ anterior cerebral

_____ anterior communicating

_____ basilar

_____ carotid

_____ middle cerebral

_____ posterior cerebral

_____ posterior communicating

_____ vertebral

C7.   Refer to Exhibit 14.4 in your textbook to answer these questions about branches off the thoracic aorta.

a. The superior phrenic artery supplies blood to the _____ and

_____ surfaces of the diaphragm.

b. The _____ _____ arteries along with the subcostal arteries are responsible for supplying the mammary glands and the vertebral canal and its contents.

c. The opening in the diaphragm for passage of the thoracic aorta is called the

_____ _____.

C8. Refer to Exhibits 14.5 and 14.6 in your textbook and Figure LG 14.4 to test your knowledge of the branches off the abdominal aorta and the principal arteries of the pelvis and lower extremities.

a. The three branches off the celiac trunk (letter _____) are the

  1.

  2.

  3.

b. The common hepatic artery gives rise to three main branches: the hepatic artery proper,

  right _____ artery, and _____ artery.

c. Which branch of the superior mesenteric artery (letter _____) supplies the pancreas and

  duodenum? _____ _____.

d. The left colic artery is a principal branch of the _____

  _____ artery (letter _____).

e. The division of the abdominal aorta into the right and left common iliac arteries (letter

  _____) occurs at the *(third? fourth? fifth?)* lumbar vertebra level.

f. The *(internal? external?)* iliac arteries (letter _____) supply the prostate gland and ductus deferens in males, and the uterus and vagina in females.

g. The external iliac arteries (letter _____) give rise to the _____ arteries,

  which become the _____ artery, at the back of the knee joint.

h. It is the *(posterior? anterior?)* tibial artery that gives rise to the peroneal artery.

i. Which lettered arteries in Figure LG 14.4 supply the following structures?

  _____ kidneys

  _____ gonads

  _____ adrenals

C9. Refer to Exhibit 14.7 in your textbook to answer these questions.

a. Deoxygenated blood returns to the right atrium via three veins. Which of these veins is

  commonly compressed during the third trimester of pregnancy? _____

  _____ _____.

b. The _____ _____ receives almost all venous blood

  from the myocardium.

C10. Refer to Exhibits 14.8 and 14.9 in your textbook to answer the following questions about venous blood flow.

a. The *(internal? external?)* jugular veins are responsible for draining blood from the sinuses deep in the brain.

b. The union of the internal jugular veins and _____ veins form the right and left brachiocephalic veins.

c. A sample of blood taken from the arm would preferably be drawn from the

_____ _____ vein.

d. The thoracic duct of the lymphatic system delivers lymph into the *(left? right?)* subclavian vein at its junction with the internal jugular. (See also Chapter 15.)

e. List the five deep veins of the upper limbs.

1. _____

2. _____

3. _____

4. _____

5. _____

C11. Review Exhibits 14.10, 14.11, and 14.12 in your textbook; then test your knowledge of the veins of the thorax, abdomen, pelvis, and lower limbs.

a. The longest vein in the body is the _____ _____ vein.

b. The _____ _____ vein receives blood from the peroneal vein.

c. Most thoracic structures are drained by a network of veins called the _____ system.

d. The inferior vena cava *(does? does not?)* receive veins from the gastrointestinal tract, spleen, pancreas, and gallbladder.

C12. The hepatic portal vein is formed by the union of the superior _____

and _____ veins.

C13. The _____ is the structure that allows for the exchanges of nutrients, gases, and wastes between the fetus and its mother.

C14. Refer to Figure 14.22 in your textbook and check your understanding of the fetal circulation. Match the structures of fetal circulation with their descriptions.

DA	ductus arteriosus	LV	ligamentum venosus
DV	ductus venosus	MUL	medial umbilical ligaments
FO	foramen ovale	UA	umbilical arteries
LA	ligamentum arteriosum	UV	umbilical vein
LT	ligamentum teres		

a. _____ Remnant of vessel that once connected pulmonary trunk and aorta.

b. _____ Vasoconstriction of the umbilical vein results in this structure.

c. _____ Atrophy of the umbilical arteries forms these.

d. _____ Passes its blood to the inferior vena cava, bypassing the liver.

e. _____ Carries blood from the fetus to the placenta.

f. _____ Permits blood to pass directly from right atrium to left atrium.

g. _____ Blood returns from the placenta to the fetus through this vessel.

h. _____ Permits blood to pass directly from the pulmonary trunk into the aorta.

i. _____ Vasoconstriction and atrophy of the ductus venosus results in formation of this structure.

D. Developmental Anatomy of Blood Vessels and Blood   (pages 471–472)

D1. _____ _____ are isolated masses and cords of mesenchyme from which blood vessels develop.

D2. Blood formation in the embryo itself begins about the *(second? third?)* month.

E. Aging and the Cardiovascular System   (page 472)

E1. Cerebral blood flow is _____ % less and renal flow is _____% less by age _____ than in the same person at age 30.

F. Applications to Health (Refer to *Applictions to Health* companion to answer Part F questions.)

F1. Match the medical term with its correct definition (you may use the answer more than once).

AN	aneurysm
DVT	deep-venous thrombosis

a. _____ A serious complication is pulmonary embolism.

b. _____ A weakened or thinned section of an artery or vein.

c. _____ The presence of a blood clot in a vein.

d. _____ Common causes for this include trauma, congenital blood vessel defects, syphilis, and atherosclerosis.

F2. Contrast primary hypertension and secondary hypertension.

F3. Name three organs commonly affected by uncontrolled hypertension.

a.

b.

c.

ANSWERS TO SELECT QUESTIONS

A1. A—endothelium, B—basement membrane, C—internal elastic lamina, D—smooth muscle, E—external elastic lamina, F—Tunica externa.

A2. Lumen.

A3. Media, elastic, smooth, elasticity, contractility.

A4. Vasoconstriction, decrease.

A5. Elastic, collagen.

A6. (a) Aorta; (b) brachiocephalic; (c) common carotid; (d) subclavian; (e) vertebral; (f) common iliac.

A7. Forward.

A8. (a) Axillary; (b) brachial; (c) radial; (d) intercostal; (e) splenic; (f) mesenteric; (g) femoral; (h) popliteal; (i) tibial.

A10. A—Muscle fiber, B—arteriole, C—metarteriole, D—precapillary sphincter, E—true capillary, F—thoroughfare channel, G—venule.

A12. Thinner.

A13. Lower.

A14. Valves.

A15. Stellate reticuloendothelial.

A16. (a) Dural venous sinus; (b) coronary sinus of the heart.

B1. (a) 5%; (b) 60%; (c) 15%.

B2. Liver, spleen.

C1. (a) A, B, C; (b) F, E; (c) H, G, D; (d) systemic capillaries of the gastrointestinal tract.

C2. Systemic circulation.

C3. (a) PC; (b) CC; (c) HPC; (d) CRC.

C4. (a) Aorta, coronary, heart, O; (b) brachiocephalic—K, common carotid—L, subclavian—M; (c) intercostal, phrenic; (d) celiac; (e) renal, superior.

C5. (a) First, largest, right, F, E; (b) axillary—G, brachial—H; (c) J, I, superficial palmar, deep palmar; (d) vertebral—D, basilar; (e) external—A; (f) internal—B, cerebral arterial circle; (g) C.

C6. A—Anterior cerebral, B—anterior communicating, C—internal carotid, D—posterior communicating, E—posterior cerebral, F—middle cerebral, G—basilar, H—vertebral.

C7. (a) Posterior, superior; (b) posterior intercostal; (c) aortic hiatus.

C8. (a) P, 1—common hepatic, 2—left gastric, 3—splenic; (b) gastric, gastroduodenal; (c) Q, inferior pancreaticoduodenal; (d) inferior mesenteric—U; (e) V, fourth; (f) internal—W; (g) X, femoral, popliteal; (h) posterior; (i) kidney—S, gonads—T, adrenals—R.

C9. (a) Inferior vena cava, (b) coronary sinus.

C10. (a) Internal; (b) subclavian; (c) median cubital; (d) left; (e) 1—radials, 2—ulnars, 3—brachials, 4—axillaries, 5—subclavians.

C11. (a) Great saphenous; (b) posterior tibial; (c) azygos; (d) does not.

C12. Mesenteric, splenic.

C13. Placenta.

C14. (a) LA; (b) LT; (c) MUL; (d) DV; (e) UA; (f) FO; (g) UV; (h) DA; (i) LV.

D1. Blood islands.

D2. Second.

E1. 20, 50, 80.

F1. (a) DVT; (b) AN; (c) DVT; (d) AN.

F3. (a) Heart; (b) brain; (c) kidneys.

SELF QUIZ

Choose the one best answer to the following questions.

1. The blood vessels that hold the largest amount of blood, at rest, are _____.

   A. capillaries
   B. veins
   C. arteries
   D. arterioles
   E. metarterioles

2. An example of a conducting artery is the

   A. axillary
   B. popliteal
   C. radial
   D. common iliac
   E. femoral

3. Which of the following statements is NOT true pertaining to distributing arteries?

   A. The tunica media contains less smooth muscle than elastic fibers.
   B. The walls are relatively thick compared to their diameter.
   C. They are capable of greater vasoconstriction and vasodilation than elastic arteries.
   D. They are also called muscular arteries.
   E. A and D are correct.

4. An anastomoses indicates a body region that

   A. is devoid of blood vessels
   B. has only one blood supply
   C. receives an alternative blood supply through a collateral branch
   D. is drained by several veins
   E. none of the above are true

5. Which structure determines whether or not blood will enter a particular capillary bed?

   A. smooth muscles of the arterioles
   B. precapillary sphincter
   C. venule smooth muscles
   D. distributing artery smooth muscle
   E. none of the above are true

6. The capillaries that have numerous microscopic pores are referred to as _____.

   A. sinusoids
   B. sinuses
   C. fenestrated
   D. continuous
   E. noncontinuous

7. The amount of blood found in the systemic veins and venules, at rest, is _____%.

   A. 60
   B. 12
   C. 9
   D. 7
   E. 2

8. All of these vessels are in the arm or hand except the

   A. axillary artery
   B. brachial artery
   C. basilic vein
   D. cephalic vein
   E. azygos vein

9. The hepatic portal circulation carries blood from the

   A. liver to the inferior vena cava
   B. aorta to the liver
   C. stomach, intestines, and spleen to the liver
   D. liver to the stomach, intestines, and spleen
   E. gallbladder to the liver

10. The structure in the fetal circulation that allows blood to travel from the pulmonary trunk to the aorta is the

    A. foramen ovale
    B. ductus venosus
    C. umbilical veins
    D. ductus arteriosus
    E. umbilical arteries

11. Which vessel, in the adult, carries oxygenated blood toward the heart?

    A. inferior vena cava
    B. aorta
    C. pulmonary veins
    D. pulmonary arteries
    E. superior vena cava

12. In the fetal circulation, the highest blood oxygen levels are found in the

    A. umbilical arteries
    B. aorta
    C. pulmonary veins
    D. umbilical veins
    E. pulmonary arteries

13. All of the following blood vessels are located in the lower limb except the

    A. dorsalis pedis
    B. small saphenous vein
    C. popliteal artery
    D. external iliac artery
    E. brachiocephalic vein

14. The _____ vein is located anterior to the vertebral column and slightly left of the midline and begins as a continuation of the left ascending lumbar vein.

    A. common iliac
    B. basilic
    C. hemiazygos
    D. hepatic portal
    E. left gonadal

15. Which of the following arteries do(es) NOT arise from the arch of the aorta?

    A. brachiocephalic
    B. right common carotid
    C. left common carotid
    D. left subclavian
    E. A and D are correct

Answer (T) True or (F) False to the following questions.

16. _____ Arteries contain valves, while veins do not.

17. _____ The exchange of nutrients and wastes occurs at the arteriole level.

18. _____ A vascular sinus has no smooth muscle to alter its diameter.

19. _____ The external carotid arteries contribute to the formation of the cerebral arterial circle.

20. _____ After birth, the umbilical vein becomes the ligamentum teres (round ligament).

Fill in the blanks.

21. _____ refers to an inflammation of a vein, often in the leg.

22. The excessive lowering of systemic blood pressure upon standing (rising) is known as _____

_____.

23. _____ refers to pain and lameness caused by defective circulation of the blood in the vessels of the limbs.

24. The tunica _____ is composed principally of elastic and collagen fibers.

25. Dilated veins caused by leaky, incompetent valves are called _____ veins.

ANSWERS TO THE SELF QUIZ

1. B	10. D	19. F
2. D	11. C	20. T
3. A	12. D	21. Phlebitis
4. C	13. E	22. Orthostatic hypotension
5. B	14. C	23. Claudication
6. C	15. B	24. Externa
7. B	16. F	25. Varicose veins
8. E	17. F	
9. C	18. T	

The Lymphatic System

CHAPTER
15

SYNOPSIS

If you were asked, on your first day of class, to name some components of the cardiovascular system, you would probably give a partially correct answer. Likewise, if you were asked a similar question pertaining to the digestive system, again you would probably answer correctly. If, however, you were called upon to name some elements of the lymphatic system it is quite possible that you would be stymied.

The lymphatic system is often considered the "forgotten" system of the body. Never receiving the amount of attention that the musculoskeletal, nervous, and cardiopulmonary systems get, the lymphatic system seems to get lost among the eleven systems of the body (overviewed in Chapter 1). Nevertheless, this lack of respect should not be misconstrued as indicative of nonimportance, for the lymphatic system is an essential component of our bodies.

Unlike the urinary system, which is located in the abdominopelvic cavity, or the respiratory system, which is localized to the head, neck, and thorax, the lymphatic system is found throughout the body.

The lymphatic system performs several vital functions. The lymphatic vessels drain the interstitial fluid that has escaped from the capillary beds, cleanse it, and return it to the general circulation. Long chain triglycerides are absorbed, via **lacteals** (lymphatic vessels), and transported away from the gastrointestinal tract to the blood. The cells associated with the lymphatic system, **B cells and T cells**, play an essential role in protecting the body from foreign invaders. Through the production of antigen-specific antibodies and T lymphocytes, the body's healthy immune system first recognizes foreign cells and chemicals, microbes, and cancer cells, then directly or indirectly destroys them.

TOPIC OUTLINE AND OBJECTIVES

A. Lymphatic Vessels and Lymph Circulation

1. List the components of the lymphatic system and their functions.
2. Describe the structure of lymphatic vessels and contrast them with veins.
3. Explain the route of lymph circulation from the lymphatic vessels into the thoracic duct and right lymphatic duct.

B. Lymphatic Tissues

4. Compare and contrast the histology, location, and functions of the thymus gland, lymph nodes, spleen, and lymphatic nodules.

C. Principal Groups of Lymph Nodes

5. Establish the location of the principal lymph nodes of the body and the areas they drain.

D. Developmental Anatomy of the Lymphatic System

E. Aging and the Immune System

6. Describe the effects of aging on the lymphatic system.

F. Applications to Health

7. Discuss the etiology and symptoms associated with acquired immune deficiency syndrome (AIDS), lymphomas, and infectious mononucleosis.

G. Key Medical Terms Associated with the Lymphatic System

SCIENTIFIC TERMINOLOGY

Find an anatomical sample word for each prefix and suffix:

Prefix/Suffix	Meaning	Sample Word
adeno-	gland	
angio-	vessel	
afferre-	to bring	
-ectomy	removal	
efferre-	to carry outward	
-itis	inflammation	
lympha-	clear water	
mega-	large	
-oma	tumor	
-stasis	to halt	

A. Lymphatic Vessels and Lymph Circulation (pages 478–482)

A1. In Chapter 12 you were first introduced to the terms lymph and lymphatic system. Check your comprehension by answering the questions below.

a. The lymphatic system consists of four principal components. List these components.

1.

2.

3.

4.

b. The primary lymphatic organs of the body are the red _____

_____ in the flat bones and epiphyses of long bones, and the

_____ gland.

A2. Briefly summarize the three principal functions of the lymphatic system.
a.

b.

c.

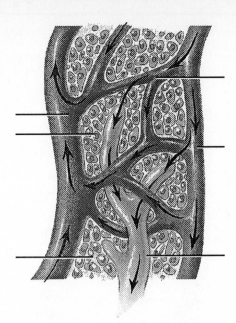

Relationship of lymphatic capillaries
to tissue cells and blood capillaries

Figure LG 15.1   Lymphatic capillaries.

A3. Refer to Figure LG 15.1 and label the following structures.

arteriole	lymph capillary
blood capillary	tissue cell
interstitial fluid	venule

A4. Lymph capillaries have a slightly *(smaller? larger?)* diameter than blood capillaries and possess a structure which permits interstitial fluid to flow *(into them but not out of them? out of them but not into them?).*

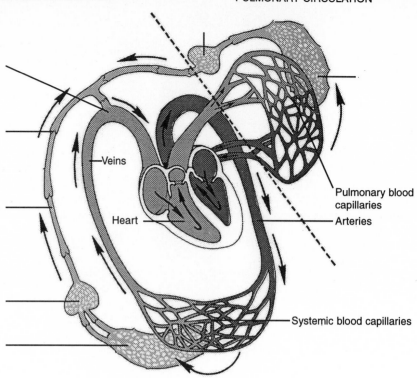

SYSTEMIC CIRCULATION

PULMONARY CIRCULATION

Veins

Heart

Pulmonary blood
capillaries

Arteries

Systemic blood capillaries

Arrows show direction of flow of lymph and blood

Figure LG 15.2   Lymphatic system.

A5. Refer to Figure 15.3 in your textbook and label the relationship of the lymphatic system
to the cardiovascular system (see Figure LG 15.2).

A6. Briefly describe how lymph is formed.

A7. The efferent vessels of the most proximal lymph nodes coalesce to form lymph trunks.
List the five principal trunks.

a. _____   d. _____

b. _____   e. _____

c. _____

A8. The two main lymph channels into which the principal trunks drain are the

_____ duct and _____ _____
duct.

A9. The dilation of the thoracic duct, located in front of the second lumbar vertebra, is called

the _____ _____.

A10. Contrast the regions drained by the two main lymph channels.

A11. Briefly describe where the thoracic duct and right lymphatic duct empty their contents.

A12. What two factors influence the flow of lymph?

a.

b.

B. Lymphatic Tissues (pages 482–486)

B1. The thymus is located in the superior _____. It is composed of two thymic lobes, which are divided into lobules, with each lobule consisting of a cortex and

a medulla. The cortex is composed of tightly packed _____, while the

medulla consists of _____ cells that produce _____ hormones.

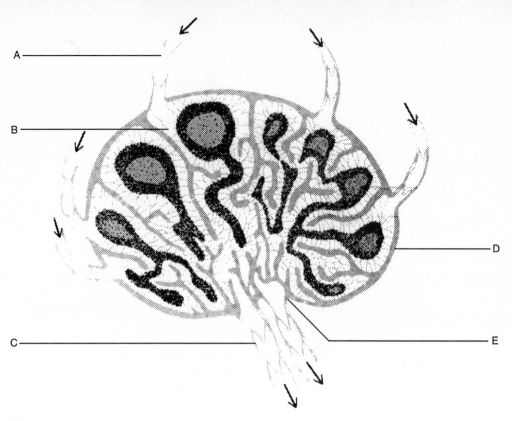

Figure LG 15.3 Structure of a lymph node.

B2. Refer to Figure LG 15.3 and test your knowledge of lymph nodes by completing the exercise below.

a. The slight depression on one side of a lymph node is referred to as the

_____. _____ lymphatic vessels emerge from the node at this point.

b. _____ are capsular extensions that penetrate into a lymph node and divide the node into compartments.

c. The two regions of a lymph node are the _____ and the

_____. The former contain many _____ , which are

regions of densely packed lymphocytes arranged in masses resembling

_____ _____ and the latter tightly packed strands

called _____ _____.

d. Lymphatic vessels which enter and exit lymph nodes *(contain? do not contain?)* valves.

e. The *(afferent? efferent?)* lymphatic vessels enter on the convex side of a lymph node.

f. Label structures A–E on Figure LG 15.3.

> _____ afferent lymphatic vessel
>
> _____ capsule
>
> _____ cortical sinus
>
> _____ efferent lymphatic vessel
>
> _____ hilus

○ color the vessels that enter the lymph node
○ color the vessels that exit the lymph node
○ site of macrophage
○ site of B lymphocyte proliferation

B3. Briefly summarize what occurs while lymph passes through a lymph node.

B4. The _____ is the largest mass of lymphatic tissue in the body.

B5. The parenchyma of the spleen consists of two different kinds of tissue:

(a) _____ _____ and (b) _____

_____.

The former is essentially lymphatic tissue consisting of (c) _____

arranged around central arteries, while the latter consists of (d) _____

_____ filled with blood and cords of splenic tissue called

(e) _____ _____.

B6. Match the tonsil with its respective location in the body (you may use an answer more than once).

> LT   lingual tonsils
> PLT palatine tonsils
> PT   pharyngeal tonsil

a. _____ Located at the base of the tongue.

b. _____ Embedded in the posterior wall of the nasopharynx.

c. _____ Situated in the tonsillar fossa between the pharyngopalatine and palatoglossal arches.

d. _____ Commonly removed in a tonsillectomy.

B7. Briefly explain the term mucosa-associated lymphoid tissue (MALT).

B8. Contrast the primary functions of the spleen and thymus.

C. Principal Groups of Lymph Nodes   (pages 486–496)

C1. Refer to Exhibits 15.1, 15.2, and 15.3; then test your knowledge of the principal lymph node of the head, neck, and upper and lower limbs. Match the lymph node(s) with their locations or the region drained.

A	axillary	P	pectoral
B	buccal	POP	popliteal
C	central	SDC	superior deep cervical
DI	deep inguinal	SI	superficial inguinal
F	facial	SM	submandibular
O	occipital	ST	supratrochlear

a. _____ Drain skin and muscles of anterior and lateral thoracic walls, and central and lateral portions of mammary glands.

b. _____ Located above medial epicondyle of humerus.

c. _____ Drain deep lymphatics of lower limb, penis, and clitoris.

d. _____ Drain posterior head and neck, auricle, tongue, larynx, esophagus, thyroid gland, and tonsils.

e. _____ Drain chin, lips, nose, nasal cavity, cheeks, gums, lower surface of palate, and anterior portion of tongue.

f. _____ Located near trapezius and semispinalis capitis muscles.

g. _____ Consist of the infraorbital, buccal, and mandibular nodes.

h. _____ Most deep lymph nodes of the upper limbs are classified as these nodes.

i. _____ Drain the knee and portions of leg and foot, especially the heel.

j. _____ Drain anterior and lateral abdominal wall, gluteal region, external genitals, perineal region, and superficial lymphatics of lower limbs.

k. _____ Drain the skin and mucous membrane of nose and cheek.

l. _____ Located at the base of axilla, embedded in adipose tissue.

C2. Why is it important to check lymph nodes when treating breast cancer?

C3. Refer to Exhibits 15.4 and 15.5 in your textbook. Test your knowledge of the lymph nodes of the abdomen and pelvis and the thorax.

a. The _____ _____ nodes drain the pelvic viscera, perineum, gluteal region, and posterior surface of the thigh.

b. The sacral nodes drain the _____, _____

_____, and posterior pelvic wall.

c. The _____ nodes drain the terminal portion of the ileum, appendix, cecum, and ascending colon.

d. Lymph nodes of the abdomen are divided into _____ lymph nodes, which are retroperitoneal, and _____ lymph nodes, which are associated with visceral arteries.

e. These nodes are located along the splenic artery: _____.

f. These nodes drain the stomach, duodenum, liver, gallbladder, and pancreas:

_____.

g. The superior mesenteric nodes are divided into _____,

_____, and _____ _____ groups.

h. The four groups that form the tracheobronchial lymph nodes are _____,

_____, _____, and _____.

i. These nodes drain the posterior part of the diaphragm: _____

_____.

D. Developmental Anatomy of the Lymphatic System   (page 496)

D1. The lymphatic system begins to develop by the end of the _____ week. The first lymph

sacs to appear are paired _____ lymph sacs.

D2. The last lymph sacs to develop are the _____ lymph sacs.

E. Aging and the Immune System   (page 496)

E1. What effect does aging have on the production of lymphocytes and antibodies?

F. Applications to Health (Refer to *Application to Health* companion to answer Part F questions.)

F1. Define the term indicator disease.

F2. The name of the *(bacterium? virus?)* responsible for AIDS is _____

_____ _____.

F3. Discuss the mechanism of infection by the AIDS virus using reverse transcription.

F4. The cell most affected by the AIDS virus is the _____ cell.

F5. List several symptoms associated with Hodgkin's disease.

What methods of treatment are available to treat it?

F6. Infectious mononucleosis is caused by the _____–

_____ virus. The ratio of affected females to males is (*2:1? 3:1?*).

ANSWERS TO SELECT QUESTIONS

A1. (a) 1—Lymph (fluid), 2—lymph vessels, 3—lymph organs or structures, 4—bone marrow; (b) Bone marrow, thymus.

A4. Larger, into them but not out of them.

A7. (a) Lumbar; (b) intestinal; (c) bronchomediastinal; (d) subclavian; (e) jugular.

A8. Thoracic, right lymphatic.

A9. Cisterna chyli.

A10. Thoracic duct receives lymph from the left side of the head, neck, and chest, the left upper limb, and the entire body below the ribs; right lymphatic duct receives lymph from the right side of the head, neck, the right upper limb, and the right side of the thorax.

A11. The thoracic duct empties into the general circulation at the junction of the left subclavian and left internal jugular veins; the right thoracic duct empties into the general circulation at the junction of the right subclavian and right internal jugular veins.

A12. (a) Skeletal muscle contraction, (b) respiratory movement.

B1. Mediastinum, lymphocytes, epithelial, thymic.

B2. (a) Hilus, efferent; (b) trabeculae; (c) cortex, medulla, follicles, lymphatic nodules, medullary cords; (d) contain; (e) afferent; (f) A—afferent lymphatic vessel, B—cortical sinus, C—efferent lymphatic vessel, D—capsule, E—hilus.

B4. Spleen.

B5. (a) White pulp; (b) red pulp; (c) lymphocytes; (d) venous sinuses; (e) splenic (Billroth's) cords.

B6. (a) LT; (b) PT; (c) PLT; (d) PLT.

C1. (a) P; (b) ST; (c) DI; (d) SDC; (e) SM; (f) O; (g) F; (h) A; (i) POP; (j) SI; (k) B; (l) C.

C3. (a) Internal iliac; (b) rectum, prostate gland; (c) ileocolic; (d) parietal, visceral; (e) pancreaticosplenic; (f) hepatic; (g) mesenteric, ileocolic, transverse mesocolic; (h) tracheal, bronchial, bronchopulmonary, pulmonary; (i) posterior phrenic.

D1. Fifth, jugular.

D2. Posterior.

F2. Virus, human immunodeficiency virus.

F4. T4.

F6. Epstein–Barr, 3:1.

SELF QUIZ

Choose the one best answer to the following questions.

1. Which of the following statements is NOT true?

   A. Lymphatic tissue not enclosed by a capsule is called diffuse lymphatic tissue.
   B. The spleen is considered a lymph organ.
   C. Lymphatic nodules have a capsule and are oval shaped.
   D. Most lymphatic nodules are small, solitary, and discrete.
   E. All of the above statements are true.

2. Which of the following is NOT a function of the lymphatic system?

   A. the lymphatic system returns protein-containing fluid (interstitial fluid) to the cardiovascular system
   B. the lymphatic system transports dietary fats from the gastrointestinal tract to the blood
   C. lymphatic tissues function in surveillance and immunological defense of the body
   D. the lymphatic system can differentiate T cells into plasma cells and secrete antibodies
   E. all of the above statements are true

3. The portion of a lymph node that contains densely packed lymphocytes is the

A. medulla
B. cortex
C. medullary cords
D. trabeculae
E. A and B are both correct

4. After exiting the afferent lymphatic vessel, the lymph enters the _____ of the lymph node.

A. medullary sinuses
B. cortical sinuses
C. efferent lymphatic vessels
D. trabeculae
E. none of the above

5. The spleen

A. phagocytizes worn-out red blood cells and platelets
B. stores and releases blood, if needed
C. serves as a site for B cell production
D. participates in blood cell formation in the fetal stage
E. all of the above are correct

6. Which of the following lymphatic structures is located in the nasopharynx?

A. adenoid
B. palatine tonsils
C. lingual tonsils
D. Peyer's patches
E. sublingual tonsils

7. If removal of the spleen is necessary, which structures in particular assume the functions of the spleen?

A. thymus gland
B. liver
C. bone marrow
D. tonsils
E. B and C

8. The cisterna chyli is located in front of the _____ lumbar vertebra.

A. first
B. second
C. third
D. fourth
E. fifth

9. The thymus gland is conspicuous in infancy and reaches a maximum size of approximately _____ grams during puberty.

A. 20
B. 30
C. 40
D. 50
E. 60

10. Which lymph node is responsible for draining the terminal portion of the ileum, appendix, cecum, and ascending colon?

A. gastric
B. hepatic
C. mesenteric
D. ileocecal
E. transverse mesocolic

11. Which of the following is NOT a primary site of lymphatic tissue?

A. spleen
B. thymus gland
C. kidneys
D. tonsils
E. lymph nodes

Answer (T) True or (F) False to the following questions.

12. _____ In a tonsillectomy, the lingual tonsils are most often removed.

13. _____ Lymph capillaries begin as closed-ended vessels.

14. _____ Lymph vessels contain no valves.

15. _____ The medullary portion of the thymus gland contains thymic corpuscles, concentric layers of epithelial cells.

16. _____ The term lymphoma refers to any tumor composed of lymphatic tissue.

17. _____ Splenomegaly refers to an enlarged spleen.

18. _____ The first lymph sacs to appear during development are the carotid lymph sacs.

19. _____ Lymph often passes through several lymph nodes prior to returning to the general circulation.

Fill in the blanks.

20. _____ is an excessive accumulation of interstitial fluid in tissue spaces, often caused by an obstruction such as an infected node.

21. The lymphatic system sometimes acts as a pathway for the movement of cancer cells throughout the body. This spread of cancer is called _____.

22. A stoppage of lymph flow is called _____.

23. _____ refers to an enlarged, tender, and inflamed lymph node resulting from an infection.

24. Aggregated lymphatic follicles in the ileum of the small intestine are called _____ patches.

25. The removal of a lymph node is called a _____.

ANSWERS TO THE SELF QUIZ

1. C	10. D	19. T
2. D	11. C	20. Edema
3. B	12. F	21. Metastasis
4. B	13. T	22. Lymphostasis
5. E	14. F	23. Adenitis
6. A	15. T	24. Peyer's
7. E	16. T	25. Lymphadenectomy
8. B	17. T	
9. C	18. F	

Nervous Tissue

CHAPTER
16

SYNOPSIS

The nervous and endocrine systems are responsible for coordinating body activities and maintaining homeostasis. As your journey continues you will first explore the wonders of the nervous system and later venture into the endocrine system.

The nervous system is divided into two principal components: the **central nervous system (CNS)** and the **peripheral nervous system (PNS).** The CNS is responsible for interpreting incoming sensory information and generating a corresponding response to the neuroeffectors (muscles and/or glands). The PNS is composed of cranial nerves and spinal nerves which are responsible for the conduction of impulses from the periphery to the CNS and from the CNS to the periphery.

Despite the complexity of the nervous system, it consists of only two principal kinds of cells: **neuroglia** *(neuro-,* nerve + *glia,* glue) and **neurons.** Neuroglia cells are specialized connective tissue cells, which far outnumber the neurons. Their functions are to isolate the neurons, provide support, and act as phagocytes. It is the responsibility of the neurons to transmit information within the nervous system. Varying in length from less than 1 mm to over 1 meter, these cells generate an electrochemical impulse that travels along its length, crossing the gaps (synapses) to other neurons, muscles, or glands.

Though often compared to a computer system or telecommunication network, the nervous system is far more exquisite in its design and operation. Bombarded each day by innumerable internal and external sensory data, the nervous system can assimilate, integrate, and eliminate or respond to this information in a manner that best maintains the body's homeostatic balance.

TOPIC OUTLINE AND OBJECTIVES

A. Nervous System Divisions

1. Describe the organization of the nervous system.

B. Histology of the Nervous System

2. Describe the structure and function of the six neuroglia of the nervous system.
3. Discuss the structure, variation, function, and classification of neurons.
4. Describe the variety of neuronal circuit patterns.

C. Applications to Health

5. Describe the process of neuron repair and the symptoms and causes of epilepsy.

SCIENTIFIC TERMINOLOGY

Find an anatomical sample word for each prefix and suffix:

Prefix/Suffix	Meaning	Sample Word
astro-	star	
auto-	self	
-cyte	cell	
dendro-	tree	
-dyma	garment	
epi-	above	
-glia	glue	
micro-	small	
neuro-	nerve	
oligo-	few	
soma-	body	

A. Nervous System Divisions   (page 501)

A1. The three basic functions the nervous system serves are _____,

_____, and _____.

A2. The *(peripheral? central?)* nervous system consists of cranial nerves and spinal nerves.

A3. Nerve impulses carried from the periphery to the central nervous system (CNS) are transmitted over *(afferent? efferent?)* neurons.

A4. The peripheral nervous system (PNS) is subdivided into the _____ and

_____ nervous systems.

A5. A motor neuron conducts impulses from the _____ to

_____ and _____.

A6. This subdivision of the autonomic nervous system speeds up the heartbeat:

_____ nervous system.

B. Histology of the Nervous System   (pages 501–511)

B1. The two principal cells associated with the nervous system are the

_____, which conduct nerve impulses, and _____, which support and protect.

B2. Test your knowledge of the neuroglia by completing the table below.

Type	Description	Function
Astrocyte		
Microglia	Small cells with few processes, derived from monocytes	
Neurolemmocytes		Produce part of the myelin sheath around each axon of a PNS neuron
Satellite cells		
Oligodendrocytes	Most common glial cells in CNS; smaller than astrocytes with fewer processes	
Ependymal cells		

B3. The three distinct parts of a neuron are the _____,

_____ _____, and _____.

B4. Check your understanding of the neuron by matching the nerve cell component with its description.

AC	axon collaterals	CS	chromatophilic substance
AH	axon hillock	DN	dendrite
AL	axolemma	LP	lipofuscin
AP	axoplasm	NF	neurofibrils
AT	axon terminals	SV	synaptic vesicles
AX	axon	SYB	synaptic end bulb

a. _____ A highly specialized projection that carries impulses away from the cell body.

b. _____ Bulb-shaped structures at the distal ends of the axon terminals.

c. _____ Cytoplasmic materials in the axon.

d. _____ Are the receiving or input portion of a neuron.

e. _____ Small cone-shaped elevation at the cell body (origin of the axon).

f. _____ Plasma membrane of the axon.

g. _____ Stores a chemical substance called a neurotransmitter.

h. _____ Yellowish brown granules; is probably related to aging.

i. _____ Composed of intermediate filaments; form the cytoskeleton, which provides support and shape for the cell.

j. _____ Side branches of axons (branch off typically at a right angle to the axon).

k. _____ Many fine processes at the distal end of an axon.

l. _____ Orderly arrangement of rough endoplasmic reticulum, the site of protein synthesis.

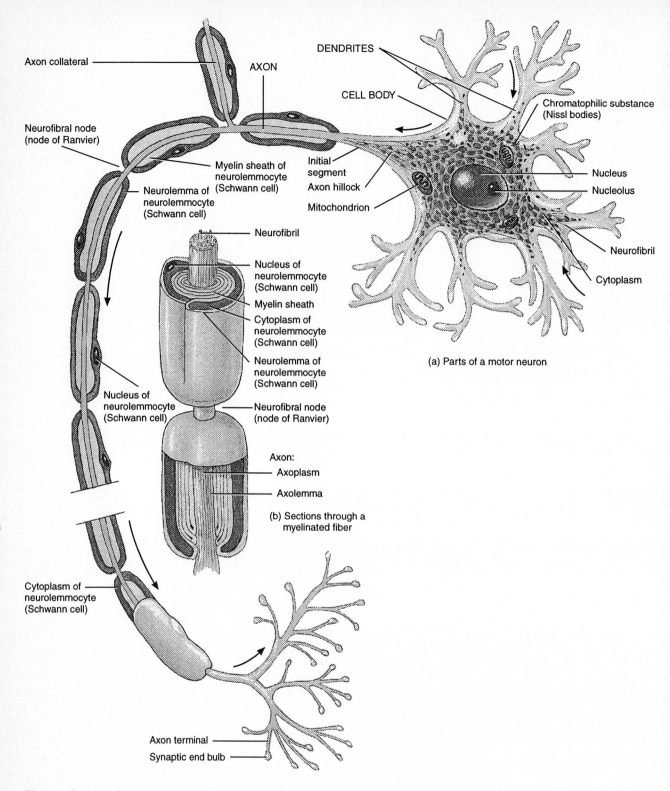

Axon collateral

AXON

DENDRITES

CELL BODY

Chromatophilic substance (Nissl bodies)

Neurofibral node (node of Ranvier)

Myelin sheath of neurolemmocyte (Schwann cell)

Initial segment

Neurolemma of neurolemmocyte (Schwann cell)

Axon hillock

Mitochondrion

Nucleus

Nucleolus

Neurofibril

Nucleus of neurolemmocyte (Schwann cell)

Myelin sheath

Cytoplasm of neurolemmocyte (Schwann cell)

Neurolemma of neurolemmocyte (Schwann cell)

Neurofibril

Cytoplasm

(a) Parts of a motor neuron

Neurofibral node (node of Ranvier)

Nucleus of neurolemmocyte (Schwann cell)

Axon:

Axoplasm

Axolemma

(b) Sections through a myelinated fiber

Cytoplasm of neurolemmocyte (Schwann cell)

Axon terminal

Synaptic end bulb

Figure LG 16.1   Structure of a typical neuron as exemplified by an efferent neuron.

B5.   Refer to Figure LG 16.1 (neuron) and answer these questions.

a.   Label all leader lines.

b.   Color in the different components indicated by the color code ovals.

B6. What is the function of the myelin sheath?

B7. A neurolemmocyte is responsible for the myelination of axons of the *(central? peripheral?)* nervous system. After wrapping itself around the axon, the remaining, outer

nucleated cytoplasmic layer is called the _____.

B8. The unmyelinated gaps between the cells that form the myelin sheath are called

_____ _____.

B9. Contrast an oligodendrocyte and a neurolemmocyte after myelination has occurred.

B10. Refer to Figure LG 16.2 and label the following neurons.

_____ bipolar

_____ multipolar

_____ unipolar

B11. Neurons that are not specifically sensory or motor neurons are called

_____ neurons. Two examples are Renshaw cells in the spinal cord and

_____ cells in the cerebellum.

B12. The processes of afferent and efferent neurons are arranged into bundles called *(tracts? nerves?)* inside the CNS.

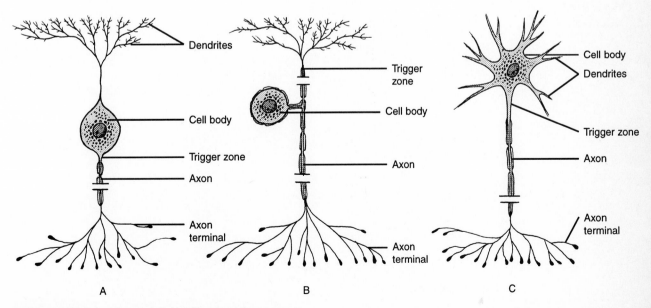

Figure LG 16.2 Structural classification of neurons.

B13. Check your understanding of nerve coverings by matching the connective tissue coat with its description.

_____ endoneurium   _____ epineurium   _____ perineurium

a. Superficial covering around the entire nerve.

b. Individual axons are wrapped by this coat.

c. This coat wraps around each fascicle.

B14. Define the following.

a. White matter

b. Gray matter

B15. List the four types of neuronal pool arrangements (circuits) common to the central nervous system.

a.

b.

c.

d.

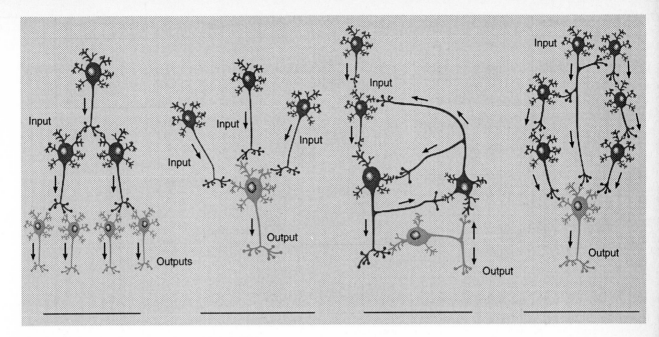

Figure LG 16.3   Examples of neuronal circuits.

B16. Refer to Figure LG 16.3 and label the diagrams as divergence, convergence, reverberating, and parallel after-discharge circuits.

C. Applications to Health (Refer to the *Applications to Health* companion to answer Part C questions.)

C1.   What is chromatolysis? Explain the results of this process.

C2.   How does retrograde degeneration differ from Wallerian degeneration?

C3.

a.   Describe the function of the neurolemma and neurolemmocytes in neuronal regeneration.

b.   Axons growing from the proximal area of damage toward the distal area of damage grow at a rate of *(1? 1.5? 2?)* mm per day.

C4.   Epilepsy is the *(most? second most?)* common neurological disorder after stroke. Epileptic seizures are initiated by *(synchronous? asynchronous?)* electrical discharges of neurons in the brain. Epilepsy almost *(never? always?)* affects intelligence.

C5.   Nerve cells lose their ability to reproduce around *(5? 6?)* months of age.

ANSWERS TO SELECT QUESTIONS

A1. Sensory, integrative, motor.
A2. Peripheral.
A3. Afferent.
A4. Somatic, autonomic.
A5. CNS, muscles, glands.
A6. Sympathetic.
B1. Neurons, neuroglia.
B2.

Type	Description	Function
Astrocyte	Star-shaped cells with processes; protoplasmic astrocytes found in gray matter of CNS, fibrous astrocytes found in white matter of CNS	Participate in the metabolism of neurotransmitters and in brain development; link neurons and blood vessels; and help form the blood–brain barrier
Microglia	Small cells with few processes, derived from monocytes	Engulf and destroy microbes and cellular debris in CNS
Neurolemmocytes	Flattened cells arranged around axons in PNS	Produce part of the myelin sheath around each axon of a PNS neuron
Satellite cells	Flattened cells arranged around the cell bodies of neurons in ganglia	Support neurons in ganglia of PNS
Oligodendrocytes	Most common glial cells in CNS; smaller than astrocytes with fewer processes	Give support to CNS neurons; produce myelin sheath around axons of CNS neurons
Ependymal cells	Epithelial cells arranged in a single layer, and range in shape from squamous to columnar; many are ciliated	Line ventricles in the brain and the central canal of the spinal cord; secrete cerebrospinal fluid and assist in circulation of CSF

B3. Dendrite, cell body, axon.
B4. (a) AX; (b) SYB; (c) AP; (d) DN; (e) AH; (f) AL; (g) SV; (h) LP; (i) NF; (j) AC; (k) AT; (l) CS.
B6. Electrically insulates the axon of a neuron and increases the speed of nerve impulse conduction.
B7. Peripheral, neurolemma.
B8. Neurofibral nodes.
B10. A—Bipolar; B—Unipolar; C—multipolar.
B11. Association, Purkinje.
B12. Tracts.
B13. A—Epineurium, B—endoneurium; C—perineurium
B15. (a) Diverging circuit; (b) converging circuit; (c) reverberating circuit; (d) parallel afterdischarge circuit.
C3. (b) 1.5.
C4. Second most, asynchronous, never.
C5. 6.

SELF QUIZ

Choose the one best answer to the following questions.

1. Which cells are responsible for the myelin sheath in the central nervous system?

   A. neurolemmocytes   C. astroglia
         D. microglia
   B. oligodendrocytes   E. satellite cells

2. Damage to the inner ear would possibly affect this type of neuron.

   A. multipolar   D. quadripolar
   B. unipolar   E. none of the
   C. bipolar   above are correct

3. Which part of a neuron is the site of protein synthesis?

   A. neurolemma   D. neurofibral
   B. neurofibrils   nodes
   C. chromatophilic   E. A and B are both
      substance   correct

4. The cells responsible for supporting neurons in ganglia in the peripheral nervous system are

   A. ependymal cells   D. astrocytes
   B. satellite cells   E. A and D are both
   C. microglia   correct

5. Which of the following is the same as a nerve fiber?

   A. a nerve, such as the median
   B. a neuron
   C. a neuroglial cell
   D. an axon or dendrite
   E. none of the above answers are correct

6. The term perikaryon refers to which part of a neuron?

   A. axon
   B. dendrite
   C. chromatophilic substance
   D. cell body
   E. axon terminal

7. _____ refers to several pre-synaptic end bulbs synapsing with a single post-synaptic neuron.

   A. divergence
   B. convergence
   C. apposition
   D. retrograde conduction
   E. C and D are both correct

8. Approximately 90% of all neurons in the body are _____ neurons.

   A. afferent
   B. motor
   C. sensory
   D. association
   E. A and C are both correct

9. The contact point between a nerve fiber and a muscle or gland is called the

   A. synapse
   B. axon hillock
   C. collateral junction
   D. ganglion
   E. tract

10. Unmyelinated gaps on a myelinated axon are called

    A. gap junctions
    B. neurofibral nodes
    C. tight junctions
    D. oligodendritic nodes
    E. C and D are both correct

11. In what part of a neuron would you find the stored neurotransmitter substance?

    A. dendrites
    B. cell body
    C. axon hillock
    D. axon terminal
    E. synaptic vesicles

Answer (T) True or (F) False to the following questions.

12. _____ Neurolemmocytes produce the myelin sheath in the peripheral nervous system.

13. _____ Microglia are phagocytic cells of nervous tissue.

14. _____ White matter refers to aggregations of myelinated processes from many neurons.

15. _____ Motor function is the ability to sense certain changes (stimuli) within the body.

16. _____ Fibrous astrocytes are found in the white matter of the CNS.

17. _____ Synaptic end bulbs are the same as side branches of the axon.

Fill in the blanks.

18. _____ neurons have only one process extending from the cell body and are always sensory neurons.

19. Purkinje and Renshaw cells are just two examples of the thousands of types of _____ neurons.

20. The output component of the CNS consists of nerve cells called _____ or efferent neurons.

21. The _____ nervous system can be divided into the somatic and autonomic nervous systems.

22. Glia outnumber neurons by _____ to _____ times.

23. The functional contact between two neurons or between a neuron and an effector is called

_____.

24. _____ is yellowish pigment contained in the cytoplasm of a neuron; it is a probable by-product of lysosomal activity.

25. The brain and spinal cord are part of the _____ nervous system.

ANSWERS TO THE SELF QUIZ

1. B	10. B	19. Association
2. B	11. E	20. Motor
3. C	12. T	21. Peripheral
4. C	13. T	22. 5, 50
5. D	14. T	23. Synapse
6. D	15. F	24. Lipofuscin
7. B	16. T	25. Central
8. D	17. F	
9. A	18. Unipolar	

The Spinal Cord and the Spinal Nerves

CHAPTER 17

SYNOPSIS

The **spinal cord** and **spinal nerves** serve as the vital link between the brain and periphery. The long ascending tracts, consisting of sensory neurons, convey nerve impulses up the spinal cord to the brain. The long descending tracts, consisting of motor neurons, conduct nerve impulses down the spinal cord, where they form neuron-to-neuron synapses and pass out with the spinal nerves.

The spinal nerves, a component of the peripheral nervous system, connect the central nervous system to **sensory receptors**, muscles, and glands.

Despite its appearance as "just" a bundle of nerve fiber tracts, forming some kind of large telephone cable, the spinal cord performs special functions that involve the processing of information on its own, such as controlling spinal reflexes. Many of these reflexes are relatively simple (monosynaptic), while other are highly complex (polysynaptic).

The spinal cord is protected and supported by the vertebral column, spinal meninges, and cerebrospinal fluid. Although it is shielded by these safeguards, the spinal cord is vulnerable to damage from a tumor or disc compression, blood clots, degenerating and demyelinating disorders, fracture or dislocation of the vertebrae, and other traumatic events. Depending upon the location and causative factors, neuralgia (*neuro* = nerve, *algia* = pain), neuritis (*itis* = inflammation), or paralysis may occur.

TOPIC OUTLINE AND OBJECTIVES

A. Spinal Cord Anatomy

1. List and describe the protective coverings of the spinal cord.
2. Discuss the general features of the spinal cord.
3. Describe the structural features found in cross-sectional examination of the spinal cord.

B. Spinal Cord Functions

4. Compare the origin, termination, and function of selected ascending and descending tracts.
5. Define the reflex center and list the basic components of a reflex arc.

C. Spinal Nerves

6. Describe the composition and covering of the spinal nerves.
7. Discuss the branches of a spinal nerve.
8. Define a plexus and list the principal plexuses and their distribution.
9. Define a dermatome and explain its correlation to spinal nerves.

D. Applications to Health

10. Discuss spinal cord injury, neuritis, sciatica, shingles, and poliomyelitis.

SCIENTIFIC TERMINOLOGY

Find an anatomical sample word for each prefix and suffix:

Prefix/Suffix	Meaning	Sample Word
arach-	spider	
di-	two	
dura-	tough	
filum-	filament	
hemi-	half	
konos-	cone	
-mater	mother	
mono-	one	
para-	beyond	
pia-	delicate	
quad-	four	

A. Spinal Cord Anatomy (pages 516–520)

A1. Complete the following questions relating to spinal cord protection and coverings.

a. In addition to the vertebral column, the spinal cord is protected by three other structures. List those structures below.

    1.

    2.

    3.

b. The _____ _____ is the outermost spinal meninx. It is

composed of _____, _____ connective tissue. It forms

a sac from the level of the _____ _____ of the occipi-

tal bone to the _____ sacral vertebra.

c. The space between the outer spinal meninx and the wall of the vertebral canal is the

_____ space. It is filled with _____ and connective
tissue.

d. The *(arachnoid? pia mater?)* is a thin, transparent, connective tissue layer that adheres to the surface of the spinal cord and brain.

e. Between the arachnoid and pia mater is the *(subdural? subarachnoid?)* space, which contains *(lymph? cerebrospinal?)* fluid.

f. Extensions of the pia mater, called _____ _____,
project laterally and fuse with the arachnoid and dura mater along the length of the cord. They protect the spinal cord against shock and sudden displacement.

A2. Test your knowledge of the general features of the spinal cord.

a. The length of the average adult spinal cord ranges from _____ to _____ cm.

b. The cervical enlargement extends from the *(C3–C7? C4–T1?)* vertebral level, while the lumbar enlargement extends from the *(T9–L1? T9–T12?)* vertebral level.

c. The tapering of the spinal cord at the L1–L2 level is known as the _____

_____. Arising from this structure is the _____

_____, an extension of the pia mater that extends inferiorly and attaches
the spinal cord to the coccyx.

d. What is the cauda equina?

A3. Refer to Figure LG 17.1; then do this activity about the spinal cord.

a. Match the letters with the corresponding structures listed below.

_____ arachnoid   _____ subarachnoid space

_____ dura mater   _____ subdural space

_____ pia mater

b. The H-shaped area in the spinal cord is formed by the *(white? gray?)* matter.

c. The lateral gray horns are present only in the _____, upper

_____, and _____ segments of the cord.

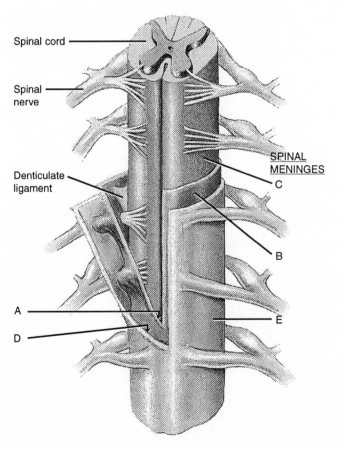

Sections through spinal cord

Figure LG 17.1   Spinal meninges.

d. The anterior, posterior, and lateral white columns consist of bundles of myelinated fibers

called tracts. The ascending spinal tracts are _____ tracts, which conduct

nerve impulses up to the brain, and the descending tracts are _____ tracts, which carry nerve impulses down the spinal cord.

B. Spinal Cord Functions   (pages 521–523)

B1. What are the primary and secondary functions of the spinal cord?

a. Primary function

b. Secondary function

B2. The spinothalamic tracts convey impulses for sensing _____,

_____, crude touch, and deep _____.

B3. The posterior column tracts carry nerve impulses for sensing the following:

a.

b.

c.

d.

e.

B4. The pyramidal tracts convey nerve impulses destined to cause precise

_____ movements of skeletal muscles. The extrapyramidal tracts

convey nerve impulses that program _____ movements, help

coordinate body movements with _____ stimuli, maintain skeletal
muscle tone and posture, and play a major role in equilibrium.

B5. The second function of the spinal cord is to serve as a center for _____

_____. The spinal nerve plays an integral role in this function and is

composed of a _____ (_____) root, which contains

sensory nerve fibers, and an _____ (_____) root,
which contains motor neurons.

B6. Each dorsal root has a swelling, the _____ (_____)

_____ ganglian, which contains the cell bodies of the

_____ neurons from the periphery.

B7. Refer to Figure LG 17.2 and complete this activity.

a. The white matter is located *(inside? outside?)* the H-shaped outline.

b. Label the parts of this figure. (Refer to Figure 17.3a in your textbook.)

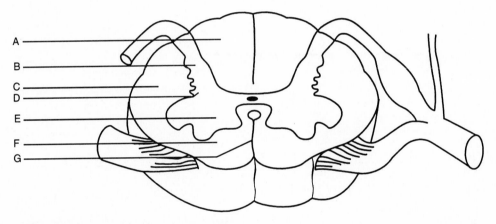

Figure LG 17.2   Spinal cord in cross section.

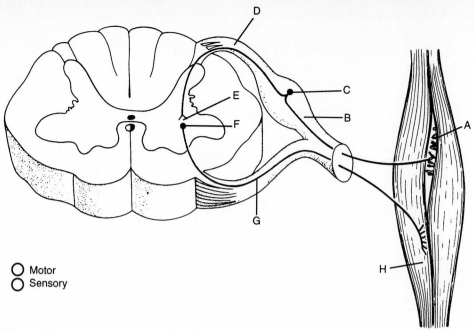

○ Motor
○ Sensory

Figure LG 17.3   Reflex arc: stretch reflex.

B8.   Refer to Figure LG 17.3 to answer these questions about the reflex arc.

a.   Label the components of a reflex arc, using the following terms. (Refer to Figure 17.5 in your textbook.)

effector	receptor
integrating center containing synapse	sensory neuron axon
motor neuron axon	sensory neuron cell body
motor neuron cell body	sensory neuron dendrite

A. _____   E. _____

B. _____   F. _____

C. _____   G. _____

D. _____   H. _____

b.   Reflexes that result in the contraction of _____ muscles are somatic reflexes. Those that result in the contraction of smooth muscle and cardiac muscle, or the

secretion of glands, are _____ reflexes.

C. Spinal Nerves   (pages 523–539)

C1.   Test your knowledge of the spinal nerve names, composition, and coverings.

a.   There are _____ pairs of spinal nerves. They are distributed as *(7? 8?)* pairs of cervical,

(11? 12? 13?) pairs of thoracic, _____ pairs of lumbar, _____ pairs of sacral, and 1 pair of coccygeal nerves.

b.   What justifies the term "mixed nerve" being applied to all spinal nerves?

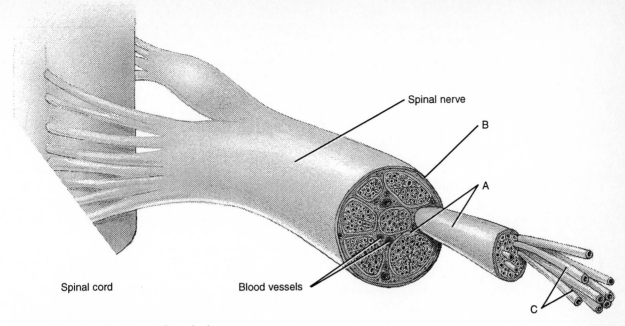

Figure LG 17.4 Coverings of a spinal nerve.

c. Individual nerve fibers, whether myelinated or unmyelinated, are wrapped in a connective

tissue called the _____, while groups of fibers are arranged in bundles
called fascicles, which are wrapped by the *(epineurium? perineurium?)*.

d. Refer to Figure LG 17.4 and insert the letter that corresponds to the following structures.
(Refer to Figure 16.4 in your textbook.)

_____ endoneurium

_____ epineurium

_____ perineurium

C2. A spinal nerve divides into four branches (rami) after exiting the intervertebral foramen.
Match the ramus with its distribution to the body.

AR	anterior ramus	RC	rami communicantes
MB	meningeal branch	PR	posterior ramus

a. _____ A component of the autonomic nervous system.

b. _____ Serves the deep muscles and skin of the dorsal trunk.

c. _____ Supplies the vertebrae, vertebral ligaments, blood vessels, and coverings of the
spinal cord.

d. _____ Serves the muscles and structures of the upper and lower extremities, and the
lateral and ventral trunk.

C3. The anterior rami of the spinal nerves, except thoracic nerves *(T1–T10? T2–T12?)*, form

a network called a _____.

C4. The cervical plexus is formed by the anterior rami of *(C1–C4? C2–C4?)*, with a contribution from C5. Complete the questions below, giving the origin and distribution of some select cervical nerves.

a. Lesser occipital

   Origin: _____

   Distribution: _____

b. Transverse cervical

   Origin: _____

   Distribution: _____

c. Ansa cervicalis (superior root)

   Origin: _____

   Distribution: _____

d. Phrenic

   Origin: _____

   Distribution: _____

C5. Match the nerve of the brachial plexus with the muscle(s) it innervates or its sensory distribution. (See Exhibit 17.2.)

AX	axillary	SS	suprascapularis
LT	long thoracic	TD	thoracodorsal
MAC	medial antebrachial cutaneous	U	ulnar
MC	musculocutaneous	US	upper subscapular
R	radial		

a. _____ Subscapularis muscle.

b. _____ Serratus anterior muscle.

c. _____ Deltoid and teres minor muscles, skin over deltoid, and upper posterior aspect of arm.

d. _____ Coracobrachialis, biceps brachii, and brachialis.

e. _____ Supraspinatus and infraspinatus muscles.

f. _____ Extensor muscles of arm and forearm; skin of posterior arm and forearm, lateral two-thirds of dorsum of hand, and fingers over the proximal and middle phalanges.

g. _____ Skin over the medial and posterior aspects of the forearm.

h. _____ Flexor carpi ulnaris and flexor digitorum profundus; skin on medial side of hand, little finger, and medial half of ring finger.

i. _____ Latissimus dorsi muscle.

C6. Check your knowledge of the distribution of the lumbar and sacral plexuses. (See Exhibits 17.4 and 17.5.)

a. The nerve that innervates the flexor muscles of the thigh is the _____ nerve. Its origin is at the *(L1–L2? L2–L4?)* level.

b. The genitofemoral nerve innervates the _____ in males and the

   _____ _____ in females.

c. The _____ nerve originates at the L2–L4 level and supplies the adductor muscles of the leg.

d. The gluteus maximus receives its neural innervation via the _____

_____ nerve.

e. The sciatic nerve (*is? is not?*) the largest nerve in the body.

f. The two nerves that compose the sciatic nerve are the _____ and

_____ _____ nerves.

C7. The ventral rami of T2–T12 (*do? do not?*) enter into the formation of plexuses. They are

known as _____ nerves.

C8. Define dermatome.

D. Applications to Health (Refer to the *Applications to Health* companion to answer Part D questions.)

D1. Match the spinal cord injury with its corresponding definition.

DP	diplegia		PP	paraplegia
MP	monoplegia		HP	hemiplegia
QP	quadriplegia			

a. _____ Paralysis of both upper or both lower extremities.

b. _____ Paralysis of one extremity only.

c. _____ Paralysis of both lower extremities.

d. _____ Paralysis of the upper extremity, trunk, and lower extremity on one side of the body.

e. _____ Paralysis of the two upper and two lower extremities.

D2. Match the following spinal cord trauma terms with their definitions.

AF	areflexia		PL	paralysis
CT	complete transection		SS	spinal shock
HS	hemisection			

a. _____ A partial transection of the spinal cord on either the right or left side.

b. _____ The loss of all reflex activity.

c. _____ Spinal cord is completely severed from one side to the other.

d. _____ Initial period following a transection. It can last from a few days to several weeks.

e. _____ The total loss of voluntary motor function that results from damage to nervous or muscle tissue.

D3. Describe some causes of neuritis.

D4. Explain the relationship of shingles and its dermatome distribution.

D5. Describe the mechanism by which the poliovirus causes paralysis.

ANSWERS TO SELECT QUESTIONS

A1. (a) 1—Meninges, 2—cerebrospinal fluid, 3—vertebral ligaments; (b) dura mater, dense, irregular, foramen magnum, second; (c) epidural, fat; (d) pia mater; (e) subarachnoid, cerebrospinal; (f) denticulate ligaments.

A2. (a) 42, 45; (b) C4–T1, T9–T12; (c) conus medullaris, filum terminale.

A3. (a) A—Subarachnoid space, B—arachnoid, C—pia mater; D—subdural space, E—dura mater. (b) gray; (c) thoracic, lumbar, sacral; (d) sensory, motor.

B1. (a) Sensory and motor nerve conduction, (b) integrating center for spinal reflexes.

B2. Pain, temperature, pressure.

B3. (a) Proprioception, awareness of the movements of muscles, tendons and joints; (b) discriminative touch, the ability to feel exactly what part of the body is touched; (c) two-point discrimination, the ability to distinguish that two different points on the skin are touched, even though close together; (d) pressure; (e) vibrations.

B4. Voluntary, automatic, visual.

B5. Spinal reflexes, posterior (dorsal), anterior (ventral).

B6. Dorsal (posterior) root, sensory.

B7. (a) Outside; (b) A—posterior columns, B—posterior horn, C—lateral columns, D—lateral horn, E—anterior horn, F—anterior columns, G—anterior median fissure.

B8. (a) A—Receptor, B—sensory neuron dendrite, C—sensory neuron cell body, D—sensory neuron axon, E—integrating center containing synapse, F—motor neuron cell body, G—motor neuron axon, H—effector; (b) skeletal, autonomic.

C1. (a) 31, 8, 12, 5, 5; (b) contain both sensory and motor fibers; (c) endoneurium, perineurium; (d) A—perineurium, B—epineurium, C—endoneurium.

C2. (a) RC; (b) PR; (c) MB; (d) AR.

C3. T2–T12; plexus.

C4. C1–C4; (a) C2, skin of scalp posterior and superior to the ear; (b) C2–C3, skin on anterior aspect of the neck; (c) C1, infrahyoid and geniohyoid muscles of the neck; (d) C3–C5, diaphragm.

C5. (a) US; (b) LT; (c) AX; (d) MC; (e) SS; (f) R; (g) MAC; (h) U; (i) TD.

C6. (a) Femoral, L2–L4; (b) scrotum, labia majora; (c) obturator; (d) inferior gluteal; (e) is; (f) tibial, common peroneal.

C7. Do not, intercostal.

D1. (a) DP; (b) MP; (c) PP; (d) HP; (e) QP.

D2. (a) HS; (b) AF; (c) CT; (d) SS; (e) PL.

SELF QUIZ

Choose the one best answer to the following questions.

1. Cerebrospinal fluid circulates in the space between the

   A. vertebrae and the dura mater
   B. dura mater and the arachnoid
   C. arachnoid and the pia mater
   D. pia mater and the spinal cord
   E. none of the above are correct

2. Which tract is classified as a pyramidal tract?

   A. rubrospinal
   B. tectospinal
   C. vestibulospinal
   D. anterior corticospinal
   E. A and C are correct

3. The inferior spinal cord tapers into a conical portion called the

   A. lumbar enlargement
   B. conus medullaris
   C. filum terminale
   D. cauda equina
   E. conus terminale

4. In the adult, the spinal cord extends from the medulla to the upper border of the _____ vertebra.

   A. seventh thoracic
   B. twelfth thoracic
   C. second lumbar
   D. first sacral
   E. fourth lumbar

5. The portion of a reflex arc that responds to the motor impulse is the

   A. receptor
   B. sensory neuron
   C. integrative center
   D. motor neuron
   E. effector

6. The spinal nerve branch that is a component of the autonomic nervous system is the

   A. posterior ramus
   B. anterior ramus
   C. meningeal branch
   D. rami communicantes
   E. none of the above answers are correct

7. Wrist drop, the inability to extend the hand at the wrist, could result from _____ nerve damage.

   A. brachial
   B. radial
   C. ulnar
   D. median
   E. A and B are correct

8. Sensory information enters the spinal cord in the

   A. posterior gray horn
   B. anterior gray horn
   C. lateral gray horn
   D. posterior white column
   E. anterior white column

9. Bundles of nerve fibers are grouped as fascicles; each fascicle is surrounded by the

   A. epineurium
   B. perineurium
   C. endoneurium
   D. endomysium
   E. perimysium

10. An injury to the _____ nerve could be indicated by an inability to flex the leg and by a loss of sensation over the anteriomedial aspect of the thigh and medial side of the leg and foot.

   A. femoral
   B. sciatic
   C. peroneal
   D. lateral femoral cutaneous
   E. iliohypogastric

Arrange the answers in the correct order.

11. From deep to superficial.   _____ _____ _____
    A. dura mater
    B. pia mater
    C. subarachnoid

12. The plexuses from superior to inferior.   _____ _____ _____ _____
    A. lumbar
    B. cervical
    C. brachial
    D. sacral

Answer (T) True or (F) False to the following questions.

13. _____ The filum terminale arises from the conus medullaris and attaches to the coccyx.

14. _____ The anterior rami of all the spinal nerves form plexuses.

15. _____ The deep, wide groove on the anterior (ventral) side of the spinal cord is called the anterior median fissure.

16. _____ Carpal tunnel syndrome involves the compression of the median nerve.

17. _____ The ascending tracts consist of sensory axons that conduct impulses up the spinal cord to the brain.

18. _____ The brachial plexus provides a nerve supply to the pelvic girdle and lower extremity.

19. _____ There are seven pairs of cervical spinal nerves.

Fill in the blanks.

20. The _____ nerve supplies the deltoid and teres minor muscles.

21. Klumpke's palsy involves the _____ roots of the _____ plexus.

22. A herniation of the disc between L5–S1 would most likely result in referred pain over the

_____ nerve.

23. The rubrospinal tract terminates in the _____ _____ _____.

24. The _____ ligaments project laterally and fuse with the arachnoid and inner surface of the dura mater.

25. Probably the most common cause of sciatica is a _____ (slipped) _____

_____.

ANSWERS TO THE SELF QUIZ

1. C
2. D
3. B
4. C
5. E
6. D
7. B
8. A
9. B

10. A
11. B, C, A
12. B, C, A, D
13. T
14. F
15. T
16. T
17. T
18. F

19. F
20. Axillary
21. Inferior, brachial
22. Sciatic
23. Anterior gray horn
24. Denticulate
25. Herniated intervertebral disc

The Brain and the Cranial Nerves

CHAPTER 18

SYNOPSIS

Your journey through the nervous system continues as our attention turns toward the most complex and wondrous organ of the body, the **brain**.

Comprising roughly 2% of the body's weight (1300 g), the brain contains about 100 billion neurons, each receiving information across as many as 80,000 synapses at one time. Functioning as the control center for the entire nervous system, the brain receives, interprets, and responds to an infinite number of nerve impulses each day. Encased in the bony cranial cavity, the brain is protected, supported, and nourished by **meninges**, **cerebrospinal fluid (CSF)**, and the **blood–brain barrier (BBB)**.

The brain is arranged in four major divisions: **brain stem, diencephalon, cerebrum,** and **cerebellum.** These regions have various functions ranging from the primitive yet vital role of regulating the respiratory and heart rates, to the refined roles of conscious perception of sensation and governance of movement.

The presence of 12 cranial nerves (along with the spinal cord—see Chapter 17) allows for the transmission of sensory and motor impulses to and from the brain.

Though unsurpassed in its design and function, the brain is not without its share of disorders. It is the brain's very design that makes it susceptible to a variety of vascular, microbial, and genetic diseases.

TOPIC OUTLINE AND OBJECTIVES

A. Brain

1. Identify the principal parts of the brain.
2. Discuss the structures that protect the brain.
3. Explain the origin, function, and circulation of cerebrospinal fluid (CSF).
4. Discuss the blood supply and the concept of the blood–brain barrier (BBB).

B. Brain Stem and Diencephalon

5. Describe the location, structural components, and function of the brain stem and diencephalon.

C. Cerebrum

6. Describe the surface features of the cerebrum.
7. List and locate the lobes of the cerebral hemisphere.
8. List the three principal types of white matter fibers underlying the cortex.
9. Compare the structure and function of the basal ganglia and limbic system.
10. Compare the sensory, motor, and association areas of the cerebrum.

D. Brain Lateralization

E. Cerebellum

11. Describe the location, structure, and functions of the cerebellum.

F. Cranial Nerves

12. Identify by name, location, type, and function the 12 pairs of cranial nerves.

G. Developmental Anatomy of the Nervous System

H. Aging and the Nervous System

I. Applications to Health

13. Describe the etiology and symptomatology of cerebrovascular accident (CVA), transient ischemic attack (TIA), Alzheimer's disease (AD), brain tumors, cerebral palsy (CP), Parkinson's disease (PD), and multiple sclerosis (MS).

J. Key Medical Terms Associated with the Central Nervous System

SCIENTIFIC TERMINOLOGY

Find an anatomical sample word for each prefix and suffix:

Prefix/Suffix	Meaning	Sample Word
a-	without	
arbor-	tree	
cauda-	tail	
chorion-	membrane	
corpus-	body	
cuneus-	wedge	
epi-	above	
-globus	ball	
glossa-	tongue	
-gnosis	knowledge	
gracila-	slender	
hemi-	half	
hydro-	water	
hypo-	under	
-kephale	head	
neuro-	nerve	
nigra-	black substance	
-phasis	speech	
proso-	before	
rhombo-	diamond shaped	
-taxes	order	
tele-	distant	
tri-	three	
vita-	life	

A. Brain   (pages 546–550)

A1.   The adult brain is divided into four principal parts. List the parts below.

a. _____

b. _____

c. _____

d. _____

A2. Recheck your knowledge of the neural coverings by listing the three meningeal layers, from superficial to deep.

a. _____

b. _____

c. _____

A3. The cranial dura consists of two layers: a thicker, outer _____ layer

and a thinner, inner _____ layer.

A4. The cerebrospinal fluid (CSF) circulates through the *(subdural? subarachnoid?)* space around the brain and spinal cord.

A5. The _____ _____ are networks of capillaries from which cerebrospinal fluid is formed by filtration and secretion. The special mechanism that prevents the passage of certain substances from the blood into the cerebrospinal fluid

is the _____ _____ _____ barrier.

A6. Refer to Figure LG 18.1 on p. 216 and complete the following questions pertaining to the divisions of the CNS and the pathway of cerebrospinal fluid around the brain and spinal cord.

a. Structures 1–3 are parts of the _____ _____.

b. Structures 4 and 5 form the _____.

c. Structure 6 represents the _____, the largest part of the brain.

d. Structure 7, the second largest part, is the _____.

e. Using the letters given in Figure LG18.1, list, in order, the pathway through which the

cerebrospinal fluid (CSF) passes. Start at the site of formation of CSF. _____ _____

_____ _____ _____ _____ _____ _____ _____ _____ _____

A7. Review the vascular components of the cerebral arterial circle (see Chapter 14). List the major arteries that supply this vascular network.

A8. The space between the penetrating blood vessels of the brain and pia mater is called a

_____ space.

A9. A condition in which cerebrospinal fluid pressure rises due to an accumulation of fluid in

the ventricles is called _____.

A10. The *(astrocytes? ependymal cells?)* are thought to pass some substances selectively from the blood into the brain but inhibit the passage of others.

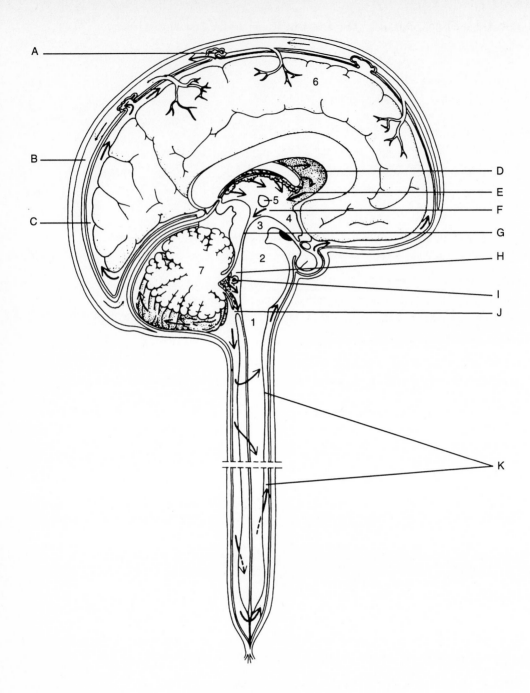

Key:

1. Medulla oblongata
2. Pons
3. Midbrain
4. Hypothalamus

5. Thalamus
6. Cerebrum
7. Cerebellum

A. Arachnoid villus
B. Cranial venous sinus
C. Subarachnoid space of brain
D. Lateral ventricle
E. Interventricular foramen
F. Third ventricle

G. Cerebral aqueduct
H. Fourth ventricle
I. Lateral aperture
J. Median aperture
K. Subarachnoid space of spinal cord

Figure LG 18.1   Brain and meninges seen in sagittal section.

B. Brain Stem and Diencephalon   (pages 551–559)

B1.   There are three parts to the brain stem. The most inferior part is the *(medulla? pons?)*,

which contains all the _____ and _____ white
matter tracts that connect the spinal cord and various parts of the brain.

B2.   Match the feature of the brain stem with the best description below.

D decussation	RAS	reticular activating system
P pyramids	RT	reticular formation
N nucleus (gracilis and cuneatus)		

a.   _____ Consists of small areas of gray matter interspersed among fibers of white matter.

b.   _____ Two anterior structures containing the largest motor tracts that pass from the outer
region of the cerebrum to the spinal cord.

c.   _____ These receive sensory fibers from ascending tracts and relay the sensory information to the thalamus on the opposite side.

d.   _____ A crossing of fibers that allows one side of the cerebrum to control the opposite
side of the body.

e.   _____ Is responsible for maintaining consciousness and wakening from sleep.

B3.   List two vital reflex centers located in the medulla.

a.

b.

B4.   The nuclei of origin for cranial nerves *(VII–XII? VIII–XI? VIII!–XII?)* are contained in the
medulla.

B5.   The oval projections on each lateral side of the medulla are called the

_____. The nuclei within these structures connect to the

_____ by paired bundles of fibers called the _____

_____ _____.

B6.   Cranial nerve VIII has an important role in *(taste? equilibrium?)*.

B7.   Test your knowledge about the pons by answering the following questions.

a.   The pons, which means "_____," is located *(posterior? superior?)* to the
medulla.

b.   The fibers of the pons run in two principal directions. The longitudinal fibers belong to the
sensory and motor tracts that connect the medulla with the upper parts of the brain stem.

The _____ fibers connect the right and left sides of the cerebellum and

form the _____ _____ _____.

c. The nuclei of these cranial nerves have their origin in the pons: CN _____, CN _____,

CN _____, and CN _____.

d. Two additional nuclei, which are located in the pons and help control respiration, are the

_____ and _____ areas.

B8. Answer the following questions about the mesencephalon (midbrain), using the abbreviations below.

CP	cerebral peduncles	RN	red nucleus
CQ	corpora quadrigemina	SC	superior colliculus
IC	inferior colliculus	SN	substantia nigra
ML	medial lemniscus	T	tectum

a. _____ Fibers from the cerebellum and cerebral cortex form synapses in this structure.

b. _____ Constitute the main connection for motor tracts between the upper parts of the brain and lower parts of the brain and the spinal cord.

c. _____ Is the posterior portion of the midbrain.

d. _____ Is a band of white fibers containing axons that convey sensory information from the medulla to the thalamus.

e. _____ Are large, darkly pigmented nuclei near the cerebral peduncles that control sub-conscious muscle activities.

f. _____ Are four rounded elevations on the posterior surface of the midbrain.

g. _____ Is a reflex center for movements of the head and trunk in response to auditory stimuli.

h. _____ Is a reflex center for movements of the eyes, head, and neck in response to visual stimuli.

B9. Which cranial nerves have their origin in the midbrain?

a. _____

b. _____

B10. List the four components of the diencephalon.

a. _____   c. _____

b. _____   d. _____

B11. The thalamus forms part of the *(walls? floor?)* of the third ventricle. It consists of *(2? 3?)* oval

masses that are joined by a bridge of gray matter called the _____

_____.

B12. Match the thalamic nuclei with the sensory impulses they transmit.

AN	anterior nucleus	VA	ventral anterior
LG	lateral geniculate	VL	ventral lateral
MG	medial geniculate	VP	ventral posterior

a. _____ voluntary motor actions and arousal (two answers)

b. _____ hearing

c. _____ vision

d. _____ certain emotions and memory

e. _____ taste, touch, pressure, vibration, temperature, and pain

B13. The thalamus comprises approximately (*70%? 80%? 90%?*) of the diencephalon.

B14. The hypothalamus is located _____ to the thalamus and is partially

protected by the _____ _____ of the sphenoid bone.

B15. The hypothalamus continually monitors both the external and internal environments. It is divided into a dozen or so nuclei in four major regions. In the questions below, match the region with its appropriate description.

M	mammillary region	S	supraoptic region
P	preoptic region	T	tuberal region

a. _____ Located anterior to the supraoptic region.

b. _____ The most posterior portion of the hypothalamus.

c. _____ Middle region, and widest portion of the hypothalamus.

d. _____ Lies above the optic chiasm.

e. _____ On its ventral surface are the tuber cinereum and infundibulum.

f. _____ Bodies within this region serve as relay stations for reflexes related to the sense of smell.

g. _____ Functions in regulating certain autonomic activities in conjunction with the hypothalamus.

h. _____ This region contains the paraventricular, supraoptic, anterior hypothalamic, and suprachiasmatic nuclei.

B16. What is the infundibulum?

B17. When sufficient food has been ingested, the _____

_____ is stimulated and sends out nerve impulses that inhibit the feeding center.

B18. The hypothalamus produces two hormones, _____ hormone and

_____, which are stored in the posterior pituitary. (See textbook Chapter 22.)

B19. Answer (T) true or (F) false to the following questions.

a. _____ The hypothalamus plays a role in regulating body temperature.

b. _____ The control of hunger and satiety are not part of the hypothalamus's function.

c. _____ The hypothalamus is one of the centers that maintain the waking state and sleep patterns.

d. _____ The hypothalamus maintains normal osmotic pressure of the extracellular fluid volume by regulating the thirst center.

e. _____ Feelings of rage and aggression are associated with the hypothalamus.

C. Cerebrum   (pages 559–567)

C1.   Test your knowledge of the cerebrum by answering the following questions.

a. The surface of the cerebrum is composed of *(gray? white?)* matter and is called the *(cortex? medulla?).*

b. Due to rapid embryonic development, the cortical region of the cerebrum rolls or folds

   upon itself. The folds are called _____ or _____. The

   deepest grooves between the folds are called _____ and the shallower

   grooves are referred to as _____.

c. The _____ fissure, which is the most prominent, separates the cerebrum into *(anterior and posterior? left and right?)* hemispheres. An extension of the cranial dura

   mater called the _____ _____ extends into this fissure.

d. The cerebral hemispheres are connected internally by a bundle of transverse white matter

   fibers called the _____ _____.

e. Refer to Figure LG 18.2 and match the following structures with the correct letters.

   _____ cerebral cortex

   _____ cerebral white matter

   _____ fissure

   _____ gyrus

   _____ sulcus

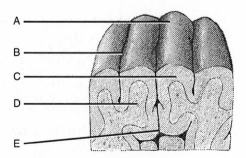

Figure LG 18.2   Cerebrum.

C2. Match the following structures with the correct description.

CS	central sulcus	PR	precentral gyrus
IS	insula	PS	postcentral gyrus
LCS	lateral cerebral sulcus	TF	transverse fissure
POS	parietooccipital sulcus		

a. _____ Lies deep within the lateral cerebral fissure, under the parietal, frontal, and temporal lobes.

b. _____ Separates the frontal and parietal lobes.

c. _____ Landmark for the primary motor area of the cerebral cortex.

d. _____ Separates the frontal lobe and temporal lobe.

e. _____ Separates the cerebrum and cerebellum.

f. _____ Separates the parietal and occipital lobes.

g. _____ Landmark for the primary somatosensory area of the cerebral cortex.

C3. List the three principal types of white matter fibers underlying the cortex. Describe the structures they connect.

a.

b.

c.

C4. Check your understanding of the basal ganglia and limbic system by answering the questions below.

a. The largest nucleus in the basal ganglia is the _____

_____.

b. The lenticular nucleus is subdivided into a lateral portion called the

_____ and a medial portion called the _____

_____.

c. The limbic system functions in emotional aspects of behavior related to

_____.

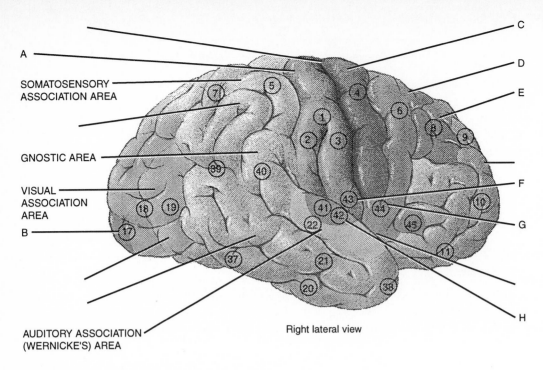

A

SOMATOSENSORY
ASSOCIATION AREA

GNOSTIC AREA

VISUAL
ASSOCIATION
AREA

B

C

D

E

F

G

H

AUDITORY ASSOCIATION
(WERNICKE'S) AREA

Right lateral view

Figure LG 18.3   Functional areas of the cerebrum.

C5.   Refer to Figure LG 18.3 to answer the following questions.

a.   Identify the following areas on the diagram.

_____	primary somatosensory	_____	primary motor
_____	primary visual	_____	premotor
_____	primary auditory	_____	frontal eye field
_____	primary gustatory	_____	Broca's area (motor speech)

b.   Using colored pencils, shade each area a different color.

C6.   *(Two? Three?)* flat electroencephalograms (complete absence of brain waves) taken *(24? 48?)* hours apart would be one criterion in establishing brain death.

C7.   Test your knowledge of the cerebrum by labeling the following structures, using the leader lines, on Figure LG 18.4.

central sulcus
frontal lobe
lateral cerebral sulcus
occipital lobe

parietal lobe
precentral gyrus
postcentral gyrus
temporal lobe

D. Brain Lateralization   (page 567)

D1.   In left-handed people the parietal and occipital lobes of the right hemisphere are usually *(wider? narrower?)* than the corresponding lobes of the left hemisphere.

E. Cerebellum   (page 568)

E1.   The extension of cranial dura mater between the cerebrum and cerebellum is called the

_____ _____.

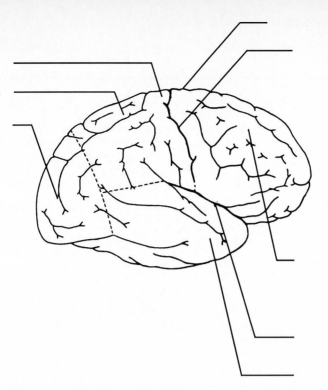

Figure LG 18.4 Right lateral view of lobes and fissures of the cerebrum.

E2. The central constriction of the cerebellum is called the _____.

E3. Between the hemispheres of the cerebellum is a sickle-shaped extension of the cranial

dura mater called the _____ _____.

E4. The _____ cerebellar peduncles contain only afferent fibers, which conduct input from the pons into the cerebellum.

F. Cranial Nerves (pages 568–581)

F1. Cranial nerves are part of the *(central? peripheral?)* nervous system. They consist of

_____ pairs with _____ originating from the brain stem.

F2. The cranial nerves are designated by roman numerals. The order in which the nerves arise is *(front to back? back to front?)*. Which is the most anterior? *(I? XII?)*

F3. Complete the table about the cranial nerves.

Number	Name	Functions
a. I		
b.	Optic	
c.		Movement of eyelid and eyeball accommodation; muscle sense; constriction of pupil
d. IV	Trigeminal	
e.		
f.		Movement of eyeball; muscle sense
g. VII		
h.	Vestibulocochlear	
i.		Saliva secretion; taste, muscle sense, blood pressure regulation
j. X		
k.	Accessory	
l.		Movement of tongue during speech and swallowing; muscle sense

F4. Complete this exercise about cranial nerves. Write the cranial nerve name and number for each description.

a. Three purely sensory cranial nerves are the _____ (_____),

   _____ (_____), and _____ (_____).

b. The largest cranial nerve, it has three branches (ophthalmic, maxillary, and mandibular):

   _____ (_____).

c. The three nerves that control eyeball movement are _____ (_____),

   _____ (_____), and _____ (_____).

d. Innervates the sternocleidomastoid and trapezius muscles: _____

   (_____).

e. Motor portion terminates in the respiratory passageways, lungs, esophagus, heart, stomach,

   and small intestine: _____ (_____).

f. Sensory portion innervates the anterior two-thirds of the tongue: _____

   (_____).

g. Injury to a branch of this nerve may cause tinnitus: _____ (_____).

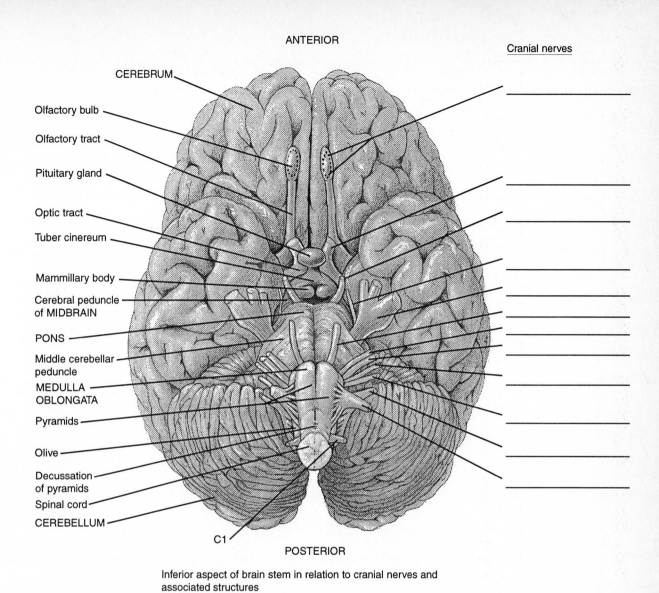

ANTERIOR

Cranial nerves

CEREBRUM

Olfactory bulb

Olfactory tract

Pituitary gland

Optic tract

Tuber cinereum

Mammillary body

Cerebral peduncle of MIDBRAIN

PONS

Middle cerebellar peduncle

MEDULLA OBLONGATA

Pyramids

Olive

Decussation of pyramids

Spinal cord

CEREBELLUM

C1

POSTERIOR

Inferior aspect of brain stem in relation to cranial nerves and associated structures

Figure LG 18.5 Brain stem.

F5. Label the cranial nerves using the leader lines on Figure LG 18.5

G. Developmental Anatomy of the Nervous System   (pages 583–585)

G1. The development of the nervous system begins early in the third week of development

with a thickening of the ectoderm called the _____

_____.

G2. List the three primary vesicles.

a.

b.

c.

G3. Two neural tube defects, _____ _____ and

_____, are associated with low levels of a B vitamin called folic acid.

H. Aging and the Nervous System   (page 586)

H1. One of the effects of aging on the nervous system is the _____

_____ _____.

H2. Conduction velocity _____, voluntary movements

_____ _____, and reflex times

_____.

H3. Impaired hearing associated with aging is known as _____.

I. Applications to Health (Refer to the *Applications to Health* companion to answer Part I questions.)

I1. A cerebrovascular accident (CVA) is more commonly known as a

_____. List three causes of CVAs.

a.

b.

c.

I2. Contrast a transient ischemic attack with a CVA.

I3. Match the name of a disorder with its definition.

AD	Alzheimer's disease	MS	multiple sclerosis
CP	cerebral palsy	PD	Parkinson's disease

a. _____ Debilitating neurological disorder of unknown origin; it afflicts about 11% of the U.S. population over age 65.

b. _____ Involves progressive degeneration of the myelin sheaths, to form hard plaques in the central nervous system.

c. _____ Degeneration of dopamine-releasing neurons in substantia nigra; results in unnecessary skeletal movement such as tremors.

d. _____ A group of motor disorders resulting in muscular incoordination and loss of muscle control.

ANSWERS TO SELECT QUESTIONS

A1. (a) Brain stem; (b) diencephalon; (c) cerebrum; (d) cerebellum.

A2. (a) Dura mater; (b) arachnoid; (c) pia mater.

A3. Periosteal, meningeal.

A4. Subarachnoid.

A5. Choroid plexuses, blood–cerebrospinal fluid.

A6. (a) Brain stem; (b) diencephalon; (c) cerebrum; (d) cerebellum; (e) D, E, F, G, H, I, J, K, C, A, B.

A7. Supplied via the right and left vertebral arteries plus the right and left internal carotid arteries.

A8. Perivascular.

A9. Hydrocephalus.

A10. Astrocytes.

B1. Medulla, ascending, descending.

B2. (a) RT; (b) P; (c) N; (d) D; (e) RAS.

B3. (a) Cardiovascular center; (b) medulbry rhythmility center.

B4. VIII–XII.

B5. Olives, cerebellum, inferior cerebellar peduncles.

B6. Equilibrium.

B7. (a) Bridge, superior; (b) transverse, middle cerebellar peduncles; (c) V, VI, VII, VIII; (d) pneumotaxic, apneustic.

B8. (a) RN; (b) CP; (c) T; (d) ML; (e) SN; (f) CQ; (g) IC; (h) SC.

B9. (a) III; (b) IV.

B10. (a) Thalamus; (b) hypothalamus; (c) epithalamus; (d) subthalamus.

B11. Walls, 2, intermediate mass.

B12. (a) VA, VL; (b) MG; (c) LG; (d) AN; (e) VP.

B13. 80%

B14. Inferior, sella turcica.

B15. (a) P; (b) M; (c) T; (d) S; (e) T; (f) M; (g) P; (h) S.

B17. Satiety center.

B18. Antidiuretic, oxytocin.

B19. (a) T; (b) F; (c) T; (d) T; (e) T.

C1. (a) Gray, cortex; (b) gyri, convolutions, fissures, sulci; (c) longitudinal, left and right, falx cerebri; (d) corpus callosum; (e) A—gyrus, B—sulcus, C—cerebral cortex, D—cerebral white matter, E—fissure.

C2. (a) IS; (b) CS; (c) PR; (d) LCS; (e) TF; (f) POS; (g) PS.

C3. (a) Association—connect gyri in same hemisphere; (b) commissural—connect corresponding gyri in opposite hemispheres; (c) projection—connect cerebrum and other parts of the brain to the spinal cord or from the spinal cord to the brain.

C4. (a) Corpus striatum; (b) putamen, globus pallidus; (c) survival.

C5. A—Primary somatosensory; B—primary visual; C—primary motor; D—premotor; E—frontal eye field; F—primary gustatory; G—Broca's area (motor speech); H—primary auditory.

C6. Two, 24.

D1. Narrower.

E1. Tentorium cerebelli.

E2. Vermis.

E3. Falx cerebelli.

E4. Middle.

F1. Peripheral, 12, 10.

F2. Front to back, I.

F3.

Number	Name	Functions
a. I	**Olfactory**	**Smell**
b. II	Optic	**Vision**
c. III	**Oculomotor**	Movement of eyelid and eyeball accommodation; muscle sense; constriction of pupil
d. IV	**Trochlear**	**Movement of eyeball, muscle sense**
e. V	Trigeminal	**Chewing, sensations of touch, pain; temperature from structures supplied, muscle sense**
f. VI	**Abducens**	Movement of eyeball; muscle sense
g. VII	**Facial**	**Facial expression, secretion of saliva and tears, muscle sense; taste**
h. VIII	Vestibulocochlear	**Hearing and equilibrium**
i. IX	**Glossopharyngeal**	Saliva secretion; taste, muscle sense, blood pressure regulation
j. X	**Vagus**	**Visceral muscle movement, sensation from organs supplied, muscle sense**
k. XI	Accessory	**Swallowing, movement of head, muscle sense**
l. XII	**Hypoglossal**	Movement of tongue during speech and swallowing; muscle sense

F4. (a) Olfactory (I), optic (II), vestibulocochlear (VIII); (b) trigeminal (V); (c) oculomotor (III), trochlear (IV), abducens (VI); (d) accessory (IX); (e) vagus (X); (f) facial (VII); (g) vestibulocochlear (VIII).

G1. Neural plate.

G2. (a) Prosencephalon; (b) mesencephalon; (c) rhombencephalon.

G3. Spina bifida, anencephaly.

H1. Loss of neurons.

H2. Decreases, slow down, increase.

H3. Presbycusis.

I1. Stroke; intracerebral hemorrhage, emboli, atherosclerosis of the cerebral arteries.

I3. (a) AD; (b) MS; (c) PD; (d) CP.

SELF QUIZ

Choose the one best answer to the following questions.

1. The inferior colliculi are associated with the

   A. thalamus
   B. pons
   C. medulla
   D. mesencephalon (midbrain)
   E. cerebral cortex

2. The nuclei for cranial nerves V, VI, VII, and part of VIII are contained within the

   A. midbrain
   B. pons
   C. diencephalon
   D. medulla
   E. cerebrum

3. Cerebrospinal fluid exits from the fourth ventricle via the

   A. interventricular foramina
   B. cerebral aqueduct
   C. lateral and median apertures
   D. arachnoid villi
   E. dural venous sinuses

4. The crossing over of motor fibers in the medulla occurs in the

   A. nucleus gracilis
   B. nucleus cuneatus
   C. decussation of pyramids
   D. inferior olive
   E. accessory olivary nuclei

5. Which nuclei of the thalamus receive sensory information pertaining to hearing?

   A. ventral lateral
   B. medial geniculate
   C. lateral geniculate
   D. ventral posterior
   E. ventral anterior

6. _____ fibers transmit impulses from the gyri in one cerebral hemisphere to the corresponding gyri in the opposite hemisphere.

   A. projection
   B. association
   C. commissural
   D. modulation
   E. none of the above answers are correct

7. The primary visual area is located in the _____ lobe.

   A. frontal
   B. parietal
   C. occipital
   D. temporal
   E. A and C are correct

8. The feeding and satiety centers are located in the _____.

   A. thalamus
   B. epithalamus
   C. pons
   D. hypothalamus
   E. cerebellum

9. The reflex control centers for heart rhythm, respiration, and blood vessel diameter are located in the

   A. cerebellum
   B. cerebrum
   C. pons
   D. midbrain
   E. medulla

10. _____ refers to the inability to carry out purposeful movements in the absence of paralysis.

   A. lethargy
   B. stupor
   C. apraxia
   D. ataxia
   E. anencephaly

Answer (T) True or (F) False to the following questions.

11. _____ The corpus callosum is a white matter bundle that connects the cerebral hemispheres.

12. _____ An inability to speak is known as aphasia.

13. _____ The caudate nucleus and lenticular nucleus are part of the basal ganglia.

14. _____ The shallow grooves between the gyri are called fissures.

15. _____ Cerebrospinal fluid is formed by the choroid plexuses in the ventricles.

16. _____ The reabsorption of cerebrospinal fluid occurs through arachnoid villi.

17. _____ The two masses of gray matter that form the thalamus are joined by the medial lemniscus.

18. _____ The white matter of the cerebrum is often called the arbor vitae.

19. _____ After exiting the optic foramen, the optic nerves unite to form the optic tracts.

Fill in the blanks.

20. The _____ _____ _____ functions in consciousness and awakening from sleep.

21. The primary motor area is located in the _____ lobe.

22. The cerebellum is separated from the cerebrum by the _____ _____ and by an extension of the cranial dura mater called the tentorium cerebelli.

23. The pneumotaxic and apneustic areas are located in the _____.

24. _____ _____ is a rare hereditary disease characterized by involuntary jerky movements and mental deterioration that terminates in dementia.

25. The inability to recognize the significance of sensory stimuli such as auditory, visual, olfactory, gustatory, and tactile stimuli is known as _____.

ANSWERS TO THE SELF QUIZ

1. D	10. C	19. F
2. B	11. T	20. Reticular activating system
3. C	12. T	21. Frontal
4. C	13. T	22. Transverse fissure
5. B	14. F	23. Pons
6. C	15. T	24. Huntington's chorea
7. C	16. T	25. Agnosia
8. D	17. F	
9. E	18. F	

General Senses and Sensory and Motor Pathways

CHAPTER 19

SYNOPSIS

As your journey in the nervous system continues, you will venture into the sensory and motor systems. In this chapter you will examine the general sense receptors and travel along the sensory and motor pathways.

Every day our nervous system is bombarded by an infinite number of internal and external stimuli. It is the function of the sensory receptors, both general and special, to convert these stimuli into nerve impulses. On the surface and within our bodies are approximately 30 million sensory neurons, each monitoring a receptor or complex of receptors.

A sensory receptor is a highly specialized structure, designed to respond to a specific stimulus rather than all stimuli. Some receptors discern specific tactile stimuli such as light touch, pressure, pain, or temperature. Other receptors respond to taste, smell, vision, hearing, or equilibrium.

After responding to a specific stimulus, the receptor transmits a nerve impulse along an ascending (sensory) pathway that carries the information to different areas within the central nervous system.

Upon receipt and interpretation of this sensory impulse, the central nervous system's response is conducted along a descending (motor) pathway to the appropriate effector.

It is this reception, transmission, interpretation, and response that allows the nervous system to be the major controller of homeostatic balance.

TOPIC OUTLINE AND OBJECTIVES

A. Sensation

1. Define sensation and discuss the levels and components of sensation.
2. Classify receptors based upon location, stimuli detected, and simplicity versus complexity.

B. General Senses

3. Identify the receptors, and their location and function, for tactile sensations (touch, pressure, and vibration), thermoreception (heat and cold), and pain.
4. List the location and function of the receptors for proprioceptive sensations.

C. Sensory (Ascending) Pathways

5. Contrast the principal pathways that convey sensory impulses from receptors to the brain.

D. Motor (Descending) Pathways

6. Compare the routes of the direct and indirect motor pathways.

SCIENTIFIC TERMINOLOGY

Find an anatomical sample word for each prefix and suffix:

Prefix/Suffix	Meaning	Sample Word
-aisthesis	sensation	
an-	without	
cuta-	skin	
kinesis-	motion	
proprio-	one's own	
tact-	touch	

A. Sensation (pages 592–593)

A1. Define the following terms.

a. Sensation

b. Perception

c. Modality

A2. For a sensation to arise, four events typically occur. List the four events below.

a.

b.

c.

d.

A3. The ability of sensory receptors to respond vigorously to one particular kind of stimulus and weakly or not at all to others would best be described by what term?

_____

A4. Match the receptor with the stimuli detected or their location.

EX exteroceptors	P proprioceptors	VR visceroceptors

a. _____ Provide information about body position and movement.

b. _____ Transmit sensations of hearing, sight, smell, touch, vibration, pressure, and pain.

c. _____ Provide information about the internal environment.

d. _____ Transmit information about muscles, tendons, joints, and the inner ear.

e. _____ Provide information about the external environment.

f. _____ Sensations arise from within the body, and often do not reach conscious perception.

A5. List the five receptors classified by the type of stimuli they detect.

a.

b.

c.

d.

A6. Simple receptors are associated with _____ senses and complex

receptors with _____ senses.

B. General Senses   (pages 593–597)

B1. Cutaneous sensations include _____ and _____
sensations, plus pain. The receptors for these sensations are *(equally? unequally?)*
distributed over the body.

B2. Tactile sensations are divided into touch, _____, and

_____, plus itch and tickle.

B3. *(True? False?)* Tactile sensations are detected by mechanoreceptors.

B4. Touch sensations include _____ touch and _____
touch. List the four receptors involved in touch sensation.

a.

b.

c.

d.

B5. The oval structures, which are layered like an onion, that detect the sensation of pressure

are the _____ _____.

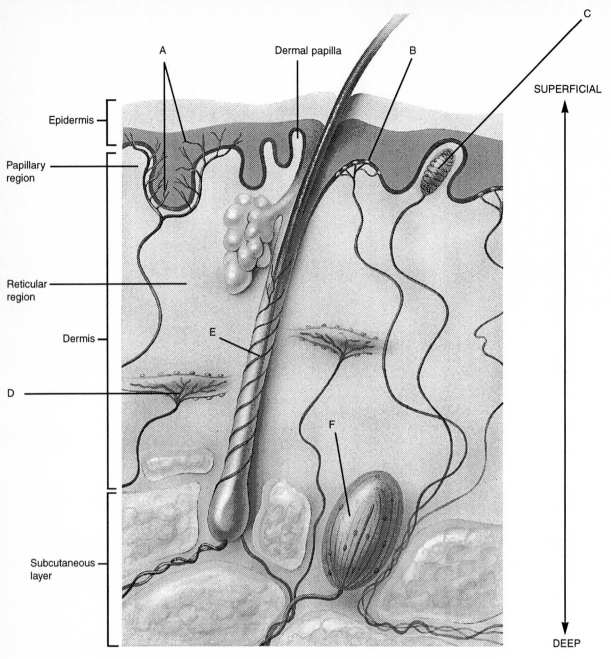

Epidermis

Papillary region

Reticular region

Dermis

D

Subcutaneous layer

A

Dermal papilla

B

C

SUPERFICIAL

E

F

DEEP

Section of skin and subcutaneous layer

Figure LG 19.1 Structure and location of cutaneous receptors.

B6. Refer to Figure LG 19.1 and match the following structures with the correct letter.

a. _____ corpuscle of touch

b. _____ hair root plexus

c. _____ lamellated corpuscle

d. _____ type I cutaneous mechanoreceptor

e. _____ type II cutaneous mechanoreceptor

f. _____ pain receptor

B7. Answer the following questions pertaining to pain sensation.

a. Another name for a pain receptor is a _____.

b. The receptors for pain are _____ _____ endings.

c. The stimulation of pain receptors in the skeletal muscles, joints, tendons, and fascia cause

   _____ _____ pain.

d. Pain that arises from stimulation of receptors in the skin is called _____

   _____ pain.

e. The sensation of pain in a surface area far from the stimulated organ is called

   _____ pain.

B8. Test your knowledge of proprioceptive sensations by matching the receptors with the corresponding description.

> JK   joint kinesthetic receptors
> MS   muscle spindles
> TO   tendon organs

a. _____ Monitor excessive tension.

b. _____ Found within and around the articular capsules of synovial joints.

c. _____ Sensitive to sudden or prolonged stretching.

d. _____ Consists of a thin capsule enclosing a few collagenous fibers, penetrated by type Ib afferent fibers.

e. _____ Monitor pressure strain on an articulation.

f. _____ Components of this receptor include the type Ia and type II fibers plus gamma motor (efferent) neurons.

C. Sensory (Ascending) Pathways   (pages 597–600)

C1. Most second-order neurons *(cross over? do not cross over?)* to the opposite side in the spinal cord and brain stem prior to ascending to the thalamus. Third-order neurons project

from the thalamus to the primary somatosensory area of the _____

cortex, which is located in the _____ gyrus.

C2. Discriminative touch, stereognosis, proprioception, weight discrimination, and vibratory

sensation involve these two tracts: fasciculi _____ and

_____ (also called the posterior column—medial lemniscus pathway).

C3. The transmission of the sensations in question C2 (above) involves three separate sensory neurons. Describe the path that the first-order, second-order, and third-order neurons take from the receptor to the cortex.

C4. The _____ spinothalamic tract conveys sensory impulses for pain and

temperature, whereas the _____ spinothalamic tract conveys impulses for tickle, itch, crude touch, and pressure.

C5. List four parts of the body that are represented by large areas in the somatosensory cortex.

a.

b.

c.

d.

C6. The posterior and anterior spinocerebellar tracts convey _____ input to the cerebellum.

D. Motor (Descending) Pathways (pages 601–606)

D1. The primary motor area is located in the _____ gyrus. After receiving and interpreting the sensory impulses, the central nervous system generates a motor response which is sent down the spinal cord in two major motor pathways, the

_____ and _____ pathways.

D2. The axons of the upper motor neurons of the pyramidal pathway terminate in the

_____ of the cranial nerves or in the _____ gray

horns of the spinal cord.

D3. The pyramidal pathway has two components: upper motor neurons located in the

_____ and spinal cord, and _____

_____ neurons which extend from the motor nuclei of nine cranial nerves to face and head muscles and the anterior horns of each spinal cord segment to

skeletal muscle fibers of the _____ and _____.

D4. Contrast the point of crossing over in the anterior and lateral corticospinal tracts.

D5. List the five major spinal cord tracts along which indirect (extrapyramidal) motor output flows and their points of origin.

a.

b.

c.

d.

ANSWERS TO SELECT QUESTIONS

A2. (a) Stimulation; (b) transduction; (c) conduction; (d) translation.
A3. Selectivity.
A4. (a) P; (b) EX; (c) VR; (d) P; (e) EX; (f) VR.
A5. (a) Mechanoreceptor; (b) thermoreceptor; (c) nociceptor; (d) photoreceptor; (e) chemo-receptor.
A6. General, special.
B1. Tactile, thermal, unequally.
B2. Pressure, vibration.
B3. True.
B4. Crude, discriminative; (a) corpuscles of touch or Meissner's corpuscles; (b) hair root plexuses; (c) tactile or Merkel discs (type I cutaneous mechanoreceptors); (d) end organs of Ruffini (type II cutaneous mechanoreceptors).
B5. Lamellated (Pacinian) corpuscles.
B6. A—Pain receptors, B—type I cutaneous mechanoreceptor, C—corpuscle of touch, D—type II cutaneous mechanoreceptor, E—hair root plexus, F—lamellated corpuscle.

B7. (a) Nociceptor; (b) free nerve; (c) deep somatic; (d) superficial somatic; (e) referred.
B8. (a) TO; (b) JK; (c) MS; (d) TO; (e) JK; (f) MS.
C1. Cross over, cerebral, postcentral.
C2. Gracilis, cuneatus.
C4. Lateral, anterior.
C5. (a) Lips; (b) face; (c) tongue; (d) thumb.
C6. Proprioceptive.
D1. Precentral, direct (pyramidal), indirect (extrapyramidal).
D2. Nuclei, anterior.
D3. Brain, lower motor, trunk, limbs.
D5. (a) Rubrospinal—origin: red nucleus of the midbrain; (b) tectospinal—origin: superior colliculus of the midbrain; (c) vestibulospinal—origin: vestibular nucleus of the medulla; (d) lateral reticulospinal—origin: reticular formation of the medulla; (e) medial reticulospinal—origin: the pons.

SELF QUIZ

Choose the one best answer to the following questions.

1. The lateral spinothalamic tract conveys information pertaining to
   A. conscious proprioception
   B. pain and temperature
   C. two-point discrimination
   D. vibration
   E. tickle

2. The receptor responsible for deep pressure is a
   A. tactile disc
   B. corpuscle of touch
   C. lamellated corpuscle
   D. nociceptor
   E. none of the above are correct

3. Which of the following does NOT belong with the others?

   A. mechanoreceptor   D. nociceptor
   B. thermoreceptor   E. proprioceptor
   C. photoreceptor

4. The ability of sensory receptors to respond vigorously to one kind of stimulus and weakly to another is termed

   A. sensation   D. selectivity
   B. perception   E. none of the
   C. conduction   above are correct

5. Spastic paralysis would be indicative of a lesion of a(n)

   A. sensory neuron   D. lower motor
   B. association   neuron
        neuron   E. A and B are both
   C. upper motor   correct
        neuron

6. Pain felt in a surface area far from the stimulated organ is called

   A. superficial pain   D. referred pain
   B. deep pain   E. somatic pain
   C. phantom pain

7. Which of the following cranial nerves is NOT part of the corticobulbar tracts?

   A. IV   D. VII
   B. V   E. VIII
   C. VI

8. With respect to the lateral corticospinal tracts, approximately _____% of the axons of upper motor neurons decussate (cross) in the medulla.

   A. 60   D. 90
   B. 70   E. 95
   C. 80

9. Which of the following tracts convey impulses on the same side of the body that regulate muscle tone in response to movements of the head?

   A. rubrospinal   D. lateral
   B. vestibulospinal   reticulospinal
   C. tectospinal   E. none of the
                    above are correct

10. Using the same answers in Question #9, choose which tract transmits impulses to the opposite side of the body to neck muscles that control movements of the head in response to visual stimuli.

11. Which of the following is NOT a function of the cerebellum?

   A. monitoring   C. comparing
       intentions   D. transduction
   B. monitoring   E. providing
       actual   corrective
       movements   feedback

12. The posterior column–medial lemniscus tract conveys information pertaining to

   A. stereognosis   D. discriminative
   B. temperature   touch
   C. crude touch   E. A and D are both
                   correct answers

13. The first-order neurons of this sensory (ascending) pathway extend from sensory receptors to the spinal cord, ultimately synapsing with second-order neurons in the medulla.
   A. anterior spinothalamic pathway
   B. posterior column-medial lemniscus pathway
   C. lateral spinothalamic pathway
   D. rubrospinal tract
   E. tectospinal tract

14. Which of the following is NOT classified as a tactile sensation?

   A. touch   D. itch
   B. heat   E. tickle
   C. pressure

15. Receptors located in blood vessels and viscera are classified as _____.

   A. exteroceptors   D. visceroceptors
   B. interoceptors   E. B and D are both
   C. proprioceptors   correct answers

Answer (T) True or (F) False to the following questions.

16. _____ The medial reticulospinal tract functions to facilitate extensor reflexes, inhibit flexor reflexes, and increase muscle tone in muscles of the axial skeleton and proximal limbs.

17. _____ Most pyramidal axons originate from cell bodies in the postcentral gyrus.

18. _____ Receptors for smell and taste are classified as complex receptors.

19. _____ Thermal sensations are included with the cutaneous sensations.

20. _____ General anesthesia blocks pain and other somatic sensations only from the point of injection downward.

ANSWERS TO THE SELF QUIZ

1. B	8. D	15. E
2. C	9. B	16. T
3. E	10. C	17. T
4. D	11. D	18. T
5. C	12. E	19. T
6. D	13. B	20. F
7. E	14. B	

Special Senses

SYNOPSIS

In the previous chapter you had the opportunity to familiarize yourself with the specialized structures known as general sensory receptors. These receptors are scattered throughout the body and are relatively simple in their design. In this chapter you will discover the highly specialized receptors for the special senses of smell, taste, vision, hearing, and equilibrium.

Unlike the general sense receptors, the special sense receptors are more complex in their structure. Functionally, however, their jobs are the same, converting stimuli into receptor or generator potentials. Relying on chemical, light,or mechanical stimuli, the receptors for special senses are some of the most interesting structures in the body.

Olfactory (smell) and **gustatory** (taste) sensations both respond to a chemical stimulation of microscopic hairs (cilia and microvilli), while visual sensation depends on light stimulating the **rods** and **cones** (photoreceptors) which are located in the retina. The sensation of hearing involves the conversion of sound waves into fluid waves via a trio of bony levers (auditory ossicles). The sense of balance requires the movement of tiny hairs covered with a jellylike substance known as the **otolithic membrane** and the **cupula**.

Let's continue our journey into the anatomy of the special sense organs and the neural pathways connecting them to the central nervous system.

TOPIC OUTLINE AND OBJECTIVES

A. Special Senses—Olfactory and Gustatory Sensations

1. Identify the receptor, location, and neural pathway for olfaction (smell).
2. Locate the gustatory receptors and describe the neural pathway for taste.

B. Special Senses—Visual Sensations

3. Describe the structural features of the eye and identify the visual pathway.

C. Special Senses—Auditory Sensations and Equilibrium

4. Describe the structural components of the ear and the auditory pathway.

5. Identify the receptors for static and dynamic equilibrium and describe the mechanism of equilibrium.

D. Applications to Health

6. Discuss the etiology and symptomatology associated with glaucoma, macular degeneration, motion sickness, deafness, Ménière's Syndrome, and otitis media.

E. Key Medical Terms Associated with Special Sensory Structures

SCIENTIFIC TERMINOLOGY

Find an anatomical sample word for each prefix and suffix:

Prefix/Suffix	Meaning	Sample Word
aqua-	water	
gusto-	taste	
kerato-	cornea	
lacrima-	tear	
laryngo-	larynx	
-lithos	stone	
-opsia	vision	
opthalmo-	eye	
photo-	light	
tympano-	drum	

A. Special Senses—Olfactory and Gustatory Sensations (pages 611–614)

A1. Answer these questions about olfactory sensation.

a. The receptors for olfactory sensation are located in the *(inferior? superior?)* portion of the nasal cavity.

b. The three principal kinds of cells of the olfactory epithelium are the

_____ _____, _____

_____, and _____ cells.

c. Olfactory cells are structurally classified as *(multipolar? unipolar? bipolar?)* neurons.

d. Cilia, called _____ _____, project from the dendrite of an olfactory cell. The cilia are the sites of olfactory transduction.

e. *(True? False?)* Mucus plays an important role in olfactory sensation.

f. The olfactory nerves (CN I) pass through multiple foramina in the _____

_____ of the ethmoid bone to enter the cranial cavity.

g. Both the supporting cells and the olfactory glands are innervated by branches of the

_____ (_____) nerve.

A2. Check your understanding of gustatory sensations by answering the questions below.

a. There are approximately _____ taste buds, found mainly on the tongue, but they are also found on the soft palate, larynx, and pharynx.

b. Each taste bud is an oval body consisting of three kinds of epithelial cells:

_____, _____, and _____ cells.

c. Gustatory hairs project to the external surface through an opening in the taste bud called a

_____ _____.

d. Taste buds are found in elevations on the tongue called papillae. Of the three types of papillae found on the tongue, the *(circumvallate? fungiform? filiform?)* papillae rarely contain taste buds.

e. List the four primary taste sensations.

1.

2. .

3.

4.

f. *(True? False?)* Gustatory and olfactory sensation are closely associated.

g. The three cranial nerves (CN) associated with gustatory sensation are CN *(II, IV, VI? VII, IX, X? V, VII, IX?).*

B. Special Senses—Visual Sensations   (page 614–623)

B1. List the five accessory structures of the eyes.

a.

b.

c.

d.

e.

B2. Answer these questions about the accessory structures of the eye.

a. The eyebrows help protect the eyeballs from _____

_____, _____, and _____

_____ _____.

b. The space between the upper and lower eyelids that exposes the eyeball is called the

_____ _____.

c. The elongated glands embedded in each tarsal plate are known as _____ glands.

d. Dilation and congestion of the blood vessels in this structure give an individual "bloodshot"

eyes: _____ _____.

e. Sebaceous glands located at the base of the hair follicles of the eyelashes are called

_____ _____ _____.

f. Briefly explain why your nose "runs" when you cry.

B3. The three tunics (layers) of the eyeball are the _____,

_____, and _____ tunics.

B4. Test your understanding of the structure of the eyeball by answering these questions.

a. The "white of the eye" is called the _____.

b. The anterior portion of the fibrous tunic is called the _____.

c. This membrane contains numerous blood vessels and a large amount of dark, brown-black

pigment and is part of the vascular tunic. Name this membrane: _____.

d. Secretion of aqueous humor is the function of the _____

_____, while alteration of the lens shape occurs via the

_____ _____. Together, these structures form the

_____ body.

e. The colored portion of the vascular tunic, suspended between the cornea and the lens, is

called the _____. The hole in the center of this structure is the

_____.

f. The retina consists of a pigmented epithelium (_____) portion and a

neural (_____) portion.

g. The three layers of the neural portion of the retina are the _____,

_____ _____, and _____

_____ layers.

B5. *(Rods? Cones?)* are responsible for color vision in bright light. The

_____ _____ is a small depression in the center of
the macula lutea which contains only cone photoreceptors.

B6. The axons of the ganglion cells extend posteriorly to the _____

_____ or blind spot. This second name is due to the *(absence?
presence?)* of rods and cones here.

B7. Answer these questions about the lens.

a. Proteins called _____, arranged like the layers of an onion, make up the lens.

b. The lens is normally *(opaque? transparent?)* and is held in place by encircling

_____ ligaments.

B8. Check your understanding of the interior of the eyeball.

a. The watery fluid located in the anterior cavity is the _____

_____, whereas the jellylike substance in the posterior cavity is called

the _____ _____.

b. Another name for the posterior cavity is the _____

_____.

c. The fluid in the anterior cavity is secreted by the _____

_____ and drains through the _____

_____ sinus.

d. Excessive intraocular pressure is called _____.

B9. Complete these questions pertaining to the visual pathways.

a. The crossing point of the optic nerves (CN II) is the _____

_____.

b. The final destination of visual impulses is the *(temporal? occipital?)* lobes of the cerebral cortex.

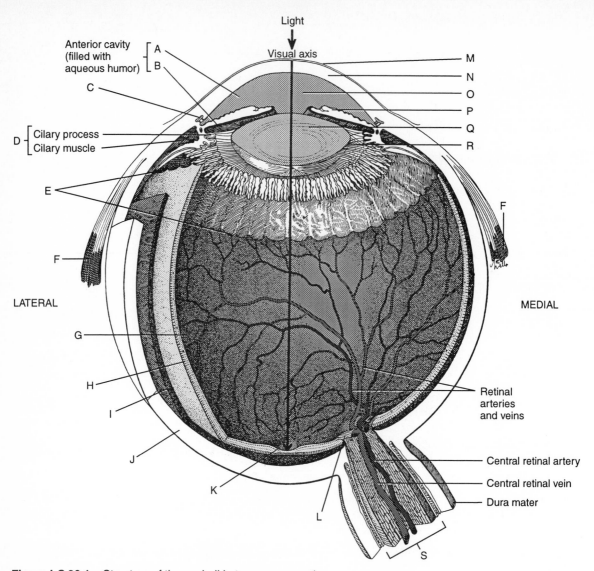

Figure LG 20.1 Structure of the eyeball in transverse section.

B10. Refer to Figure LG 20.1 to answer the following questions.

a. Match the following structures with the corresponding letter.

_____ anterior chamber

_____ bulbar conjunctiva

_____ central fovea

_____ choroid

_____ ciliary body

_____ cornea

_____ extrinsic eye muscle

_____ iris

_____ lens

_____ optic disc

_____ optic nerve

_____ ora serrata

_____ posterior chamber

_____ pupil

_____ retina

_____ sclera

_____ scleral venous sinus

_____ suspensory ligament

_____ vitreous chamber

b. Write the letters of labeled structures that form each layer.

fibrous tunic _____ _____

vascular tunic _____ _____ _____

C. Special Senses—Auditory Sensations and Equilibrium   (pages 623–634)

C1. The following questions relate to the external and middle ear.

a. The _____ is an elastic cartilage flap which is shaped like the flared end of a trumpet and is covered by thick skin.

b. The _____ _____ _____ leads from the auricle to the eardrum. This tube contains specialized sebaceous glands called

_____ glands.

c. The thin, semitransparent partition of fibrous connective tissue between the external

auditory canal and the middle ear is the _____ _____.

d. Name the three auditory ossicles: _____, _____, and

_____.

e. Two structures of interest in the middle ear include the _____

_____, which communicates with the mastoid air cells of the temporal

bone, and the _____ _____, which connects the middle ear with the nasopharynx of the throat.

f. The stapes fits into a membrane-covered opening called the _____

_____.

g. Two skeletal muscles associated with the auditory ossicles are the _____

_____ and _____ muscles.

C2. Answer these questions about the internal ear by using the abbreviations below.

A	ampulla	PL	perilymph
BL	bony labyrinth	SC	semicircular canals
EL	endolymph	US	utricle and saccule
ML	membranous labyrinth	V	vestibule

a. _____ Three bony passages, each arranged at approximately right angles to the other two.

b. _____ A series of cavities in the petrous portion of the temporal bone.

c. _____ Fluid found in the membranous labyrinth.

d. _____ Fluid similar to cerebrospinal fluid that surrounds the membranous labyrinth.

e. _____ A series of sacs and tubes lying inside, and having the same general form as, the bony labyrinth.

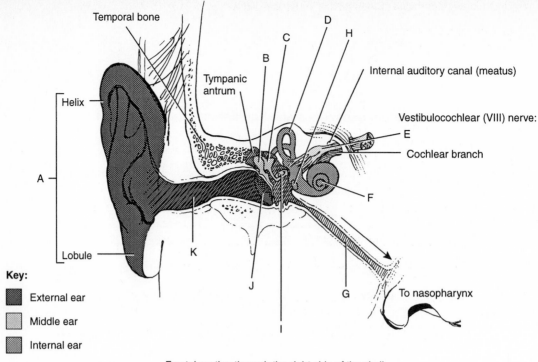

Temporal bone

Tympanic
antrum

Helix

A

Lobule

K

J

C

B

D

H

Internal auditory canal (meatus)

Vestibulocochlear (VIII) nerve:

E

Cochlear branch

F

G

To nasopharynx

I

Key:

■ External ear

□ Middle ear

▨ Internal ear

Frontal section through the right side of the skull
showing the three principal regions of the ear

Figure LG 20.2   Diagram of the ear in frontal section.

f.   _____ The oval, central portion of the bony labyrinth.

g.   _____ Two sacs of the membranous labyrinth in the vestibule.

h.   _____ A swelling at the end of each of the semicircular canals.

C3.   Lying in front of (anterior to) the vestibule is the _____, which
resembles a snail's shell.

C4.   Refer to Figure LG 20.2 and answer these questions.

a.   Color the three parts of the ear, using the color code ovals.

   ○   external ear
   ○   middle ear
   ○   inner ear

b.   Using the leader lines, label each lettered structure.

C5.   Answer these questions about the auditory pathway.

a.   Sound waves striking the tympanic membrane cause the *(incus? malleus?)* to vibrate.

b.   The movement of the stapes pushes the membrane of the *(round? oval?)* window in and
out.

c.   The *(perilymph? endolymph?)* of the scala vestibuli is pushed by the bulging of the oval
window.

d.   The hair cells of the spiral organ move against the *(basilar? tectorial?)* membrane.

e.   Auditory nerve impulses travel over the *(vestibular? cochlear?)* portion of CN *(VII? VIII?
IX?)*.

C6. Check your knowledge of the mechanism of equilibrium by answering these questions.

a. Define the following terms.
   1. Static equilibrium

   2. Dynamic equilibrium

b. The receptors for static equilibrium are the _____, which are located in

   the _____ and _____.

c. Hair cells have long extensions of the cell membrane consisting of 70 or more

   _____ and one _____ anchored firmly to its basal
   body and extending beyond the longest microvilli.

d. The thick, gelatinous glycoprotein layer floating directly over the hair cells of the macula is

   called the _____ _____.

e. The nerve impulses for equilibrium are carried on the _____ branch of
   CN *(VII? VIII?)*.

f. The receptors for dynamic equilibrium, which are located in the ampullae of the semicircular

   ducts, are called the _____. This positioning permits the detection of

   rotational _____ or _____.

g. The gelatinous mass covering the hair cells of each crista is called the

   _____.

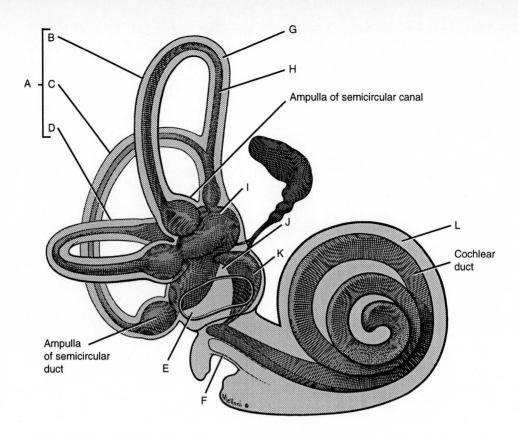

Figure LG 20.3   The internal ear.

C7. Refer to Figure LG 20.3 and match following structures with the corresponding letter.

_____ anterior semicircular canal   _____ posterior semicircular canal

_____ bony labyrinth   _____ round window

_____ cochlea   _____ saccule

_____ lateral semicircular canal   _____ semicircular canals

_____ membranous labyrinth   _____ utricles

_____ oval window   _____ vestibule

D. Applications to Health (Refer to the *Applications to Health* companion to answer Part D questions.)

D1.   What is glaucoma? Discuss the mechanism by which glaucoma may cause blindness.

D2. Test your knowledge of these visual, auditory, and vestibular disorders.

CD	conduction deafness	OM	otitis media
MO	motion sickness	SMD	senile macular degeneration
MS	Ménière's syndrome	SND	sensorineural deafness

a. _____ Nausea and vomiting brought on by repetitive angular, linear, or vertical motion. The cause is excessive stimulation of the vestibular apparatus by motion.

b. _____ The growth of new blood vessels or hard mass under the retina, leading to distorted vision or blindness.

c. _____ An acute infection of the middle ear, caused primarily by bacteria.

d. _____ Lack of the sense of hearing caused by impairment of the cochlea or cochlear branch of cranial nerve VIII.

e. _____ An increased amount of endolymph that enlarges the membranous labyrinth, causing fluctuating hearing loss, attacks of vertigo, and roaring tinnitus.

f. _____ Lack of the sense of hearing caused by an impairment of the external and middle ear mechanisms for transmitting sounds to the cochlea.

ANSWERS TO SELECT QUESTIONS

A1. (a) Superior; (b) supporting cells, olfactory receptors, basal; (c) bipolar; (d) olfactory hairs; (e) true; (f) cribriform plate; (g) facial (VII).

A2. (a) 10,000; (b) supporting, gustatory, basal; (c) taste pore; (d) filiform; (e) salty, sweet, sour, bitter; (f) true; (g) VII, IX, X.

B1. (a) Eyelids; (b) eyelashes; (c) eyebrows; (d) lacrimal apparatus; (e) extrinsic eye muscles.

B2. (a) Foreign objects, perspiration, direct sun rays; (b) palpebral fissure; (c) tarsal (Meibomian); (d) bulbar conjunctiva; (e) sebaceous ciliary glands or glands of Zeis.

B3. Fibrous, vascular, retinal (nervous).

B4. (a) Sclera; (b) cornea; (c) choroid; (d) ciliary processes, ciliary muscle, ciliary; (e) iris, pupil; (f) nonvisual, visual; (g) photoreceptor, bipolar cell, ganglion cell.

B5. Cones, central fovea.

B6. Optic disc, absence.

B7. (a) Crystallins; (b) transparent, suspensory.

B8. (a) Aqueous humor, vitreous body; (b) vitreous chamber; (c) ciliary processes, scleral venous; (d) glaucoma.

B9. (a) Optic chiasma; (b) occipital.

B10. (a) A—Anterior chamber, B—posterior chamber, C—scleral venous sinus, D—ciliary body, E—ora serrata, F—extrinsic eye muscle, G—vitreous chamber, H—retina, I—choroid, J—sclera, K—central fovea, L—optic disc, M—bulbar conjunctiva, N—cornea, O—pupil, P—iris, Q—lens, R—suspensory ligament, S—optic nerve; (b) fibrous tunic: N, J; vascular tunic: I, D, P.

C1. (a) Auricle; (b) external auditory canal (meatus), ceruminous; (c) tympanic membrane (eardrum), (d) malleus, incus, stapes; (e) tympanic antrum, auditory (Eustachian) tube; (f) oval window; (g) tensor tympani, stapedius.

C2. (a) SC; (b) BL; (c) EL; (d) PL; (e) ML; (f) V; (g) US; (h) A.

C3. Cochlea.

C4. (b) A—Auricle, B—malleus, C—incus, D—semicircular canals, E—vestibular branch, F—cochlea, G—auditory tube, H—round window, I—stapes in oval window, J—tympanic membrane, K—external auditory canal.

C5. (a) Malleus; (b) oval; (c) perilymph; (d) tectorial; (e) cochlear, VIII.

C6. (b) Maculae, utricle, saccule; (c) stereocilia, kinocilium; (d) otolithic membrane; (e) vestibular, VIII; (f) cristae, acceleration, deceleration; (g) cupula.

C7. A—Semicircular canals, B—anterior semicircular canal, C—posterior semicircular canal, D—lateral semicircular canal, E—oval window, F—round window, G—bony labyrinth, H—membranous labyrinth, I—utricle, J—vestibule, K—saccule, L—cochlea.

D2. (a) MO; (b) SMD; (c) OM; (d) SND; (e) MS; (f) CD.

SELF QUIZ

Choose the one best answer to the following questions.

1. Sebaceous glands at the base of the hair follicles of the eyelashes are called

   A. tarsal glands
   B. lacrimal glands
   C. ciliary glands
   D. ceruminous gland
   E. none of the above are correct

2. Each taste bud contains approximately _____ gustatory (taste) receptor cells.

   A. 20
   B. 30
   C. 40
   D. 50
   E. 60

3. Which cranial nerve is involved in the process of tearing when exposed to certain irritant odors?

   A. V
   B. VI
   C. VII
   D. VIII

4. This nonvascular structure is a part of the fibrous tunic.

   A. sclera
   B. ciliary body
   C. iris
   D. cornea
   E. conjunctiva

5. The maculae and cristae are classified as

   A. nociceptors
   B. proprioceptors
   C. viscero-receptors
   D. mechano-receptors
   E. B and D are correct

6. Which lingual papillae rarely contain taste buds?

   A. filiform
   B. fungiform
   C. circumvallate
   D. all of the above papillae have taste buds
   E. A and C are correct

7. The white portion of the eyeball is called the

   A. conjunctiva
   B. sclera
   C. choroid
   D. uvea
   E. cornea

8. The ciliary body is part of the

   A. fibrous tunic
   B. vascular tunic
   C. nervous tunic
   D. retinal tunic
   E. none of the above are correct

9. Which of the following is NOT part of the hearing mechanism?

   A. malleus
   B. tectorial membrane
   C. scala vestibuli
   D. tympanic membrane
   E. macula lutea

10. Information about equilibrium is transmitted by which cranial nerve?

    A. I
    B. II
    C. VII
    D. VIII
    E. IX

11. The gelatinous mass called the cupula is associated with the

    A. muscle spindle
    B. macula
    C. crista
    D. nociceptor
    E. joint kinesthetic receptor

12. The posterior wall of the middle ear communicates with the _____ through the tympanic antrum.

    A. auditory tube
    B. round window
    C. oval window
    D. mastoid air cells
    E. none of the above answers are correct

13. Some axons of cranial nerve II cross over at the _____.

    A. optic nerve
    B. optic tract
    C. optic chiasma
    D. optic bulb
    E. A and D are both correct

14. The _____ is the site where the optic nerve exits the eyeball.

    A. central fovea
    B. macula lutea
    C. optic disc
    D. ora serrata
    E. both A and C are correct

15. The scleral venous sinus drains aqueous humor from the _____.

    A. vitreous chamber
    B. posterior chamber
    C. anterior chamber
    D. posterior cavity
    E. none of the above are correct

Answer (T) True or (F) False to the following questions.

16. _____ The thick, gelatinous glycoprotein layer that rests on the hair cells of the macular is called the otolithic membrane.

17. _____ The maculae are the receptors for static equilibrium.

18. _____ There are approximately 50–60 million receptors for olfactory sense located in the nasal epithelium.

19. _____ Static equilibrium is the maintenance of body position in response to sudden movements.

Fill in the blanks.

20. _____ refers to an abnormal visual intolerance to light.

21. The movement of the stapes pushes on the membrane covering the _____ window.

22. The receptors for color vision and sharpness of vision are the _____.

23. _____ _____ is secreted into the posterior chamber and reabsorbed through the scleral venous sinus.

24. Gustatory sensation involves these three cranial nerves: _____, _____, and

    _____.

25. The scientific term for a pain receptor is a _____.

26. _____ refers to a pathological process that may be hereditary in which new bone is deposited around the oval window.

27. The turning outward of the eyes is called _____.

28. _____ refers to a ringing, roaring, or clicking in the ears.

ANSWERS TO THE SELF QUIZ

1. C	11. C	21. Oval
2. D	12. D	22. Cones
3. C	13. C	23. Aqueous humor
4. D	14. C	24. Facial, glossopharyngeal, vagus
5. E	15. C	25. Nociceptor
6. A	16. T	26. Otosclerosis
7. B	17. T	27. Exotropia
8. B	18. F	28. Tinnitus
9. E	19. F	
10. D	20. Photophobia	

The Autonomic Nervous System

CHAPTER
21

SYNOPSIS

Imagine having to concentrate each morning on digesting your breakfast or having to focus on increasing your heart rate when climbing the stairs. Luckily, these functions and many others are looked after by a special branch of the nervous system.

The **autonomic nervous system (ANS)** continually and without conscious contrast coordinates visceral activities, regulating cardiovascular, digestive, endocrine, reproductive, and respiratory functions. The regulation of these activities is essential to the maintenance of homeostasis.

Monitoring any changes in the internal and external environment, the autonomic nervous system, by controlling the visceral system, can adjust the body's water, nutrient, blood gas, and electrolyte levels to preserve homeostatic balance.

The term autonomic was originally assigned because it was thought that this system functioned completely independent of the central nervous system. Anatomists and physiologists have revised this view, for it is now known that the ANS is intimately connected to the cerebral cortex, the medulla oblongata, and the hypothalamus.

The ANS consists of two principal divisions: the **sympathetic** and the **parasympathetic.** Most organs have a dual innervation, receiving impulses from both divisions. One division transmits impulses that excite or activate an organ, while the other decreases the organ's activity. It is this delicate balancing act, carried out by these two divisions, that allows for instantaneous adjustments to be made to our systems.

TOPIC OUTLINE AND OBJECTIVES

A. Comparison of Somatic and Autonomic Nervous Systems

1. Compare the structure and function of the autonomic nervous system to the somatic efferent nervous system.

B. Anatomy of Autonomic Motor Pathways

2. Identify the components of the autonomic nervous system.
3. Contrast the location, structural features, and function of the sympathetic and parasympathetic divisions.

C. Activities of the Autonomic Nervous System

4. Identify the principal neurotransmitters of the autonomic nervous system.
5. Describe the receptors involved in autonomic response.

D. Autonomic Reflexes

6. Define an autonomic reflex and discuss the components necessary to its function.

E. Control by Higher Centers

7. Discuss the importance of the higher central nervous system centers to autonomic control.

Comparison of Somatic and Autonomic Nervous Systems   (page 640)

A1. The somatic efferent nervous system produces skeletal muscle movement (voluntary), while the autonomic nervous system regulates _____ activities.

A2. List the three types of visceral effector tissues.

a.

b.

c.

A3. In the somatic efferent pathway there is _____ motor neuron between the CNS and the effector, whereas the autonomic motor pathway consists of _____ motor neurons in series.

A4. The two principal divisions of the ANS are:

a.

b.

c. The fact that most viscera receive impulses from both divisions is referred to as

_____ _____.

B. Anatomy of Autonomic Motor Pathways   (pages 641–647)

B1. The first of the two autonomic motor neurons is called a _____ neuron. Its cell body is in the_____ or _____ _____ and it has a(n) *(myelinated? unmyelinated?)* axon.

B2. The postganglionic neuron lies entirely _____ the CNS. Its cell body and dendrites are located in an _____ _____.

B3. Answer these questions about the pre- and postganglionic neurons.

a. The preganglionic neurons of the sympathetic division have their cell bodies located in the *(anterior? lateral?)* gray horns of the first thoracic to the _____ _____ lumbar segments of the spinal cord. For this reason it is also referred to as the _____ division.

b. The cell bodies for the parasympathetic preganglionic neurons are found in the nuclei of cranial nerves _____, _____, _____, and _____, and in the lateral gray horns of the _____ through _____ sacral segments. Hence its alternate name, the _____ division.

B4. Complete the table below pertaining to the three general groups of autonomic ganglia. Give their location and the division that innervates them.

a.

Name	Location	Supplied by
1. Sympathetic trunk		
2.	Anterior to the vertebral column and close to the large abdominal arteries	
3.		Parasympathetic

b. The preganglionic neurons of the sympathetic division synapse with _____ or more

postganglionic fibers, whereas those of the parasympathetic division synapse with only

_____ to _____ postsynaptic neurons.

B5. Test your knowledge of the sympathetic and parasympathetic divisions of the ANS.

a. After exiting through the intervertebral foramina, the myelinated preganglionic sympathetic fibers enter a short pathway called a *(white? gray?)* ramus before passing to the nearest *(sympathetic trunk? prevertebral?)* ganglion on the same side.

b. Typically, there are *(15–18? 22–25?)* ganglia in each sympathetic trunk (chain). Give the typical number of ganglia associated with each vertebral region.

       _____ cervical   _____ lumbar

       _____ thoracic   _____ sacral

c. Name the three cervical ganglia and give their locations.

   1.

   2.

   3.

d. The postganglionic fibers from the thoracic sympathetic trunk innervate the

_____, _____, _____, and other

thoracic viscera. In the skin, they also innervate _____

_____, _____ _____, and arrector
pili muscles of hair follicles.

e. After entering a sympathetic trunk ganglion, the preganglionic fibers may synapse with a postganglion fiber at the level of entry, or above or below this level. The

_____ _____ _____ is the structure
containing the postganglionic fibers that connect the ganglion to the spinal nerve.

f. Some preganglionic sympathetic fibers pass through the sympathetic trunk without termi-

nating in the trunk. Beyond the trunk they form nerves called _____

nerves. These nerves from the thoracic area terminate in the _____
ganglion.

g. List the four pairs of ganglia supplied by the cranial portion of the parasympathetic division.

1. _____   3. _____

2. _____   4. _____

h. Part of the cranial outflow of the preganglionic fibers leave the brain as part of the

_____ (X) nerves. These nerves carry nearly _____% of the total
craniosacral outflow.

i. The ganglia associated with the cranial outflow are *(close to? far from?)* their visceral
effectors.

B6. In order to check your understanding of the sympathetic pathway, refer to Figure
LG 21.1. The preganglionic neuron at the fifth thoracic level has its axon drawn as far as
the white ramus. Complete the pathways to the stomach and to a sweat gland in the skin
of the thoracic region. When performing this exercise, use solid lines to indicate the
preganglionic fibers and dotted lines to indicate the postganglionic fibers.

C. Activities of the Autonomic Nervous System   (pages 648–652)

C1. Based upon their neurotransmitter, autonomic fibers are classified as either cholinergic
(release acetylcholine) or adrenergic (release norepinephrine). Using the abbreviations
below, answer the following questions dealing with neurotransmitters.

C cholinergic	A adrenergic

a. _____ Most sympathetic postganglionic axons.

b. _____ All sympathetic and parasympathetic preganglionic axons.

c. _____ Effects are longer lasting and more widespread.

d. _____ All parasympathetic postganglionic axons.

e. _____ A few sympathetic postganglionic axons.

f. _____ Effects are short-lived and localized.

C2. The two types of cholinergic (ACh) postsynaptic receptors are the

_____ and _____ receptors, whereas the two
adrenergic receptors for norepinephrine and epinephrine are called

_____ and _____ receptors.

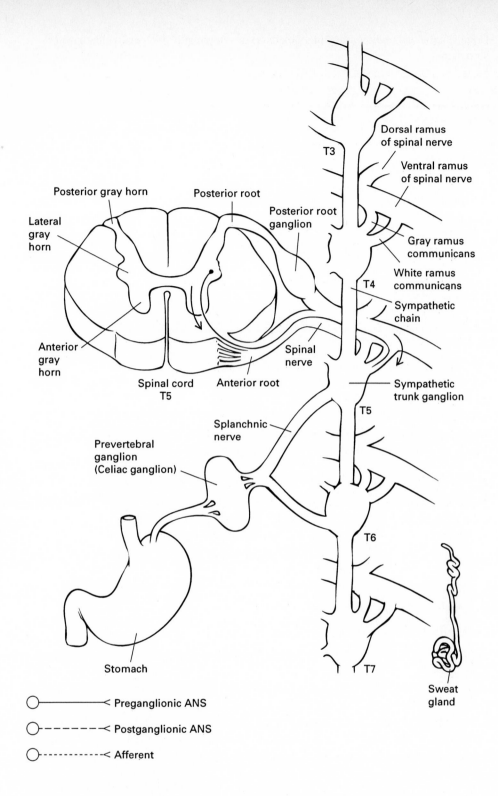

Figure LG 21.1   Typical sympathetic pathway beginning at level T5 of the cord.

Posterior gray horn

Posterior root

Posterior root ganglion

Lateral gray horn

Dorsal ramus of spinal nerve

Ventral ramus of spinal nerve

Gray ramus communicans

White ramus communicans

Sympathetic chain

Spinal nerve

Sympathetic trunk ganglion

Anterior gray horn

Spinal cord T5

Anterior root

Splanchnic nerve

Prevertebral ganglion (Celiac ganglion)

Stomach

Sweat gland

T3

T4

T5

T6

T7

○————< Preganglionic ANS

○------< Postganglionic ANS

○----------< Afferent

C3. Complete the table below, showing the effects of sympathetic and parasympathetic stimulation.

Effector	Sympathetic	Parasympathetic
a. Sweat glands		
b. Lacrimal glands		
c. Ciliary muscle of eye		
d. Lungs, bronchial muscle		
e. Stomach and intestines		
f. Kidney		
g. Gallbladder		
h. Heart arterioles		

C4. The sympathetic division is concerned with activities that _____ energy while the parasympathetic is concerned with processes involving the

_____ of energy.

C5. When confronted with a stressful situation, the sympathetic division starts a series of

physiological responses called the _____ or _____ response.

D. Autonomic Reflexes   (pages 652–653)

D1. List the five components of a visceral autonomic reflex.

a.

b.

c.

d.

e.

D2. *(All? Some?)* visceral sensations remain at a subconscious level.

E. Control by Higher Centers   (page 653)

E1. Many parts of the CNS connect with the ANS. Output from the _____ influences autonomic centers in the medulla and spinal cord.

E2. Control of the ANS by the _____ _____ occurs primarily during emotional stress.

ANSWERS TO SELECT QUESTIONS

A1. Visceral.

A2. (a) Smooth muscle; (b) cardiac muscle; (c) glands.

A3. One, two.

A4. (a) Sympathetic; (b) parasympathetic; (c) dual innervation.

B1. Preganglionic, brain, spinal cord, myelinated.

B2. Outside, autonomic ganglion.

B3. (a) Lateral, first two or three, thoracolumbar; (b) III, VII, IX, X, S2, S4, craniosacral.

B4. (a)

Name	Location	Supplied by
1. Sympathetic trunk	Vertical row on either side of the spine	Sympathetic division
2. **Prevertebral ganglion**	Anterior to the vertebral column and close to the large abdominal arteries	Sympathetic division
3. **Terminal ganglion**	Close to or actually within a visceral effectors	Parasympathetic division

(b) 20, four, five

B5. (a) White, sympathetic trunk; (b) 22–25, 3 cervical, 11–12 thoracic, 4–5 lumbar, 4–5 sacral; (c) superior—posterior to the internal carotid arteries and anterior to the transverse processes of C2, middle—near C6 at the level of the cricoid cartilage, inferior—near the first ribs, anterior to the transverse processes of C7; (d) heart, lungs, bronchi, sweat glands, blood vessels; (e) gray ramus communicans; (f) splanchnic, celiac;

(g) ciliary, pterygopalatine, submandibular, otic; (h) vagus, 80; (i) close to.

Use Figure 21.4 from your textbook to check your answers.

C1. (a) A; (b) C; (c) A; (d) C; (e) C; (f) C.

C2. Nicotinic, muscarinic, alpha, beta.

C3.

Effector	Sympathetic	Parasympathetic
a. Sweat glands	Stimulate secretion on the palms and soles and most body regions	None
b. Lacrimal glands	None	Stimulates secretion
c. Ciliary muscle of eye	Relaxation for far vision	Contraction for near vision
d. Lungs, bronchial muscle	Relaxation: airway dilation	Contraction: airway constriction
e. Stomach and intestines	Decreases motility, contracts sphincters	Increases motility, relaxes sphincters
f. Kidney	Constriction of blood vessels, decrease urine output	None
g. Gallbladder	Relaxation	Contraction
h. Heart arterioles	Relaxation → dilation	Contraction → constriction

C4. Expend, conservation.

C5. Fight, flight.

D1. (a) Receptor; (b) sensory neuron; (c) integrating center; (d) two motor neurons; (e) visceral effector.

D2. Some.

E1. Hypothalamus.

E2. Cerebral cortex.

SELF QUIZ

Choose the one best answer to the following questions.

1. Which of the following structures is NOT under autonomic nervous system control?

   A. smooth muscle contraction
   B. glandular secretion
   C. cardiac muscle contraction
   D. skeletal muscle contraction
   E. none of the above are correct

2. The preganglionic neurons of the sympathetic division have their cell bodies in (the)

   A. cranial nerves III, VII, IX, X
   B. lateral gray horns of the 12 thoracic segments
   C. lateral gray horns of the first two or three lumbar segments
   D. lateral gray horns of the second through fourth sacral segments
   E. A and D are correct answers
   F. B and C are correct answers

3. Which of the following responses is NOT associated with the parasympathetic division?

   A. stimulates secretion of the gastric glands
   B. decreases the motility of the ureter
   C. increases the rate and force of heart contraction
   D. causes pupillary constriction of the eye
   E. causes bronchial airway constriction

4. The ganglia associated with the parasympathetic division are the

   A. vertebral chain
   B. paravertebral
   C. collateral
   D. prevertebral
   E. terminal

5. The sympathetic division is primarily involved with activities that

   A. restore the body's energy reserve
   B. expend the body's energy
   C. conserve the body's energy
   D. occur when not under stress
   E. A and D are correct answers

6. Typically there are _____ vertebral chain ganglia.

   A. 3
   B. 4–6
   C. 11–15
   D. 22–25
   E. 26–30

7. The pterygopalatine ganglion supplies all the following structures except

   A. smooth muscle of the eyes
   B. nasal mucosa
   C. palate
   D. pharynx
   E. lacrimal apparatus

8. All preganglionic neurons are

   A. sympathetic
   B. unmyelinated
   C. parasympathetic
   D. myelinated
   E. A and B are both correct answers

Answer (T) True or (F) False to the following questions.

9. _____ Sympathetic stimulation to the spleen causes relaxation and increased storage of blood.

10. _____ All postganglionic sympathetic fibers are cholinergic.

11. _____ All viscera have a dual innervation from the sympathetic and parasympathetic divisions of the ANS.

12. _____ The white rami communicantes connect the sympathetic trunk to the spinal nerves.

13. _____ The visceral efferent pathway of the ANS almost always consists of two efferent neurons.

14. _____ The sympathetic division is distributed throughout the body, including the skin.

15. _____ Parasympathetic postganglionic fibers run to terminal ganglia.

Fill in the blanks.

16. Sympathetic stimulation of the kidneys leads to _____ of the blood vessels and results in decreased urine volume.

17. _____ receptors are found on all effectors (muscles and glands) innervated by parasympathetic postganglionic axons.

18. Cholinergic postganglionic axons release _____.

19. Increased respiratory and heart rates, plus increased blood levels of glucose and epinephrine, would occur as a result of _____ stimulation.

20. The major control and integration center of the ANS is the _____.

21. The acronym "SLUD" is a mental key for remembering many parasympathetic responses. It stands for salivation, _____, urination, and _____.

22. The division of the ANS that is totally cholinergic is the _____.

ANSWERS TO THE SELF QUIZ

1. D
2. F
3. C
4. E
5. B
6. D
7. A
8. D

9. F
10. F
11. F
12. F
13. T
14. T
15. F
16. Constriction

17. Muscarinic
18. Acetylcholine
19. Sympathetic
20. Hypothalamus
21. Lacrimation, defecation
22. Parasympathetic

The Endocrine System

SYNOPSIS

There are two systems involved in the control of body function and the maintenance of home-ostasis: the **nervous system** and the **endocrine system.**

In the last six chapters you have examined the wonders of the nervous system, gaining insight into its role and function in the body. Your voyage through the body now turns to the endocrine system. You will examine the location, histology, and blood and nerve supply of the nine principal endocrine glands, plus other endocrine tissues. A discussion of the etiology and symptomatology of disorders associated with each gland is included.

The nervous system controls homeostasis by reacting almost instantaneously and convey-ing electrical impulses along neurons to specific effectors. The endocrine system affects body function and homeostasis by releasing **hormones** (chemical messengers) directly into the bloodstream.

Whereas neurons exert their effects quickly, hormones may take up to several hours to affect a response. In addition, the nervous system's response tends to generate a localized response, while the endocrine system sends a message to cells throughout the body.

It is the release of these chemical messengers that allows the endocrine system to con-tribute to the basic processes of life, including the regulation of (1) metabolism and energy bal-ance, (2) growth and development, (3) the internal environment's chemical composition and volume, (4) certain components of the immune system, and (5) basic reproductive activities.

In addition, the endocrine system helps the body cope with stresses such as infection, trauma, dehydration, starvation, temperature extremes, and hemorrhage.

TOPIC OUTLINE AND OBJECTIVES

A. Endocrine Glands and Overview of Hor-mone Effects

1. Define an endocrine and an exocrine gland.
2. List the endocrine glands of the body.

B. Hypothalamus and Pituitary Gland (Hypophysis)

3. Contrast the structure and function of the adenohypophysis and neurohypophysis.
4. Describe the location, histology, and blood and nerve supply of the pituitary gland.

C. Thyroid Gland

5. Describe the location, histology, and blood and nerve supply of the thyroid gland.

D. Parathyroid Glands

6. Describe the location, histology, and blood and nerve supply of the parathyroid glands.

E. Adrenal (Suprarenal) Glands

7. Describe the location, histology, and blood and nerve supply of the adrenal glands.
8. Contrast the cortical and medullary subdivisions of the adrenal glands.

F. Pancreas

9. Describe the location, histology, and blood and nerve supply of the pancreas.

G. Ovaries and Testes

10. List the hormones secreted by the ovaries and testes.

H. Pineal Gland (Epiphysis Cerebri)

11. Describe the location, histology, and blood and nerve supply of the pineal gland.

I. Thymus Gland

12. Describe the location, histology, and blood and nerve supply of the thymus gland.

J. Other Endocrine Tissues

13. List other endocrine tissues and their hormones.

K. Developmental Anatomy of the Endocrine System

L. Aging and the Endocrine System

SCIENTIFIC TERMINOLOGY

Find an anatomical sample word for each prefix and suffix:

Prefix/Suffix	Meaning	Sample Word
acro-	limb	
adeno-	glandular	
endo-	within	
ex-	out	
gonas-	seed	
gyneco-	woman	
hormon-	to urge on	
lact-	milk	
meli-	honey	
myxa-	mucus	
pan-	all	
soma-	body	

A. Endocrine Glands and Overview of Hormone Effects   (pages 658–659)

A1. The two regulatory systems of the body are the _____ and the

_____. The former controls homeostasis via neurons, while the latter

releases chemical messenger molecules, called _____.

A2. Refer to Figure 22.1 in your textbook and briefly list seven functions of hormones.

a.

b.

c.

d.

e.

f.

g.

A3. Exocrine glands secrete their products into _____ or to the outer

_____ of the body, whereas the endocrine glands release their

secretions into the extracellular space and then into the _____.

A4. List the five (*exclusive*) endocrine glands named in the textbook.

a.

b.

c.

d.

e.

A5. A given hormone will travel throughout the body via the circulatory system. It affects

only specific cells called _____ cells. These cells are influenced by
binding of hormones to large protein or glycoprotein molecules called

_____.

A6. If a hormone is present in excess, the number of target cell receptors may (*decrease?
increase?*).

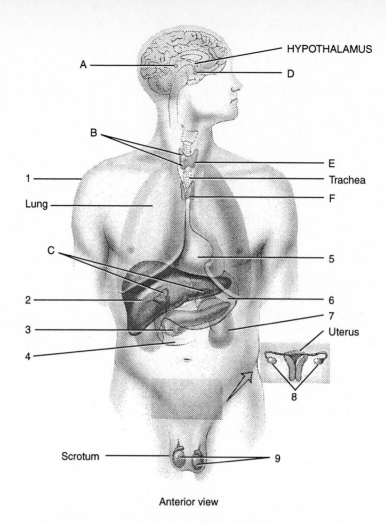

Figure LG 22.1   Diagram of endocrine glands and organs containing endocrine tissue.

A7. Refer to Figure LG 22.1 and complete the following exercise.

a. Color, then label, the endocrine glands A–F.

b. Label each organ (numbers 1–9) that contains endocrine tissue.

B. Hypothalamus and Pituitary Gland (Hypophysis)   (pages 660–664)

B1. The pituitary has the nickname the "_____" gland because it secretes several hormones that control other endocrine glands. Located in the sella turcica, the pituitary gland is *(anterior? posterior?)* to the optic chiasma.

B2. The *(pars intermedia? infundibulum?)* is a stalklike structure that attaches the pituitary gland to the hypothalamus.

B3. The pituitary gland is divided structurally and functionally into an anterior lobe called the

_____ and a posterior lobe referred to as the _____.

B4. Release of anterior pituitary hormones is stimulated by _____

_____ and suppressed by _____

_____ from the hypothalamus.

B5. Match the glandular cell with its secretion.

> CLC   corticotroph cells
> GC   gonadotroph cells
> LC   lactotroph cells
> SC   somatotroph cells
> TC   thyrotroph cells

a.   _____ human growth hormone (hGH)

b.   _____ luteinizing hormone (LH)

c.   _____ thyroid-stimulating hormone (TSH)

d.   _____ prolactin (PRL)

e.   _____ melanocyte-stimulating hormone (MSH)

f.   _____ follicle-stimulating hormone (FSH)

g.   _____ adrenocorticotropic hormone (ACTH)

B6. A _____ hormone is one that influences another endocrine gland.

B7. The neurohypophysis does not synthesize hormones; rather, it stores hormones produced

by _____ cells of the hypothalamus. The two hormones stored in the

posterior pituitary gland are _____ hormone and

_____.

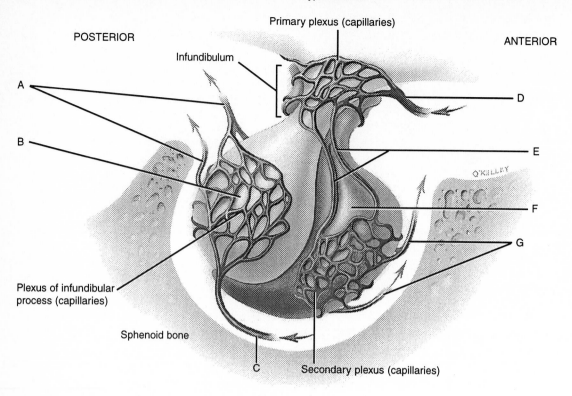

Figure LG 22.2   Blood supply of the pituitary gland.

B8. Refer to Figure LG 22.2 and match the structures below to the corresponding letter..

_____ anterior hypophyseal vein   _____ posterior hypophyseal vein

_____ anterior pituitary   _____ posterior pituitary

_____ hypophyseal portal veins   _____ superior hypophyseal artery

_____ inferior hypophyseal artery

C. Thyroid Gland   (pages 664–666)

C1. The thyroid gland is located inferior to the larynx and has two lobes connected by a mass

called the _____. The gland has an excellent blood supply, receiving

about _____ to _____ ml of blood per minute.

C2. Name the two types of cells associated with the thyroid follicles and the hormones they secrete.

Cell	Hormone(s)
a.	
b.	

C3. Contrast the actions of the hormones listed in question C2.

D. Parathyroid Glands (page 666)

D1. The parathyroid glands are attached to the *(anterior? posterior?)* surfaces of the lateral lobes of the thyroid gland. The two types of cells contained in the parathyroid glands are

_____ and _____ cells.

D2. Parathyroid hormone (PTH) is responsible for *(decreasing? increasing?)* the blood *(K+ and Na+? Ca2+ and Mg2+?)* levels, while *(decreasing? increasing?)* the blood *(phosphate? bicarbonate?)* levels.

E. Adrenal (Suprarenal) Glands (pages 666–669)

E1. The adrenal glands are structurally and functionally divided into two regions: the outer

_____ _____ and the inner _____

_____.

E2. The outer region of the adrenal gland is subdivided into three zones. Complete the table below by matching the zone with its hormone grouping.

Zone	Hormone Grouping
a. Glomerulosa	
b. Fasciculata	
c.	Gonadocorticoids

E3. The cells of the adrenal medulla are called _____ cells. These cells are directly innervated by preganglionic cells of the *(parasympathetic? sympathetic?)* division of the ANS.

E4. The two principal hormones synthesized by the adrenal medulla are

_____ and _____. Both of these hormones are sympathomimetic. What does that mean?

F. Pancreas (pages 670–671)

F1. The pancreas is both an _____ and _____ gland.

F2. The endocrine portion of the pancreas consists of cell clusters called

_____ _____. These clusters are composed of four kinds of cells. Match the cell with its secretion and its principal effect.

A alpha cells	B beta cells
D delta cells	F F-cells

a. _____ Secrete pancreatic polypeptide.

b. _____ Secrete the hormone glucagon.

c. _____ Secrete somatostatin.

d. _____ Secrete the hormone insulin.

e. _____ Regulate the release of pancreatic digestive enzymes.

f. _____ Lower the blood sugar level.

g. _____ Raise the blood sugar level.

h. _____ Inhibit the secretion of insulin and glucagon.

F3. The exocrine portion of the pancreas is formed by clusters of cells called

_____ , which are responsible for the secretion of digestive enzymes.

F4. Pancreatic secretion is controlled largely by the hormones _____ and

_____ released by the small intestine.

G. Ovaries and Testes   (page 672)

G1. The ovaries are the female gonads and the testes are the male gonads. Complete the table below pertaining to their hormones and actions.

Hormone	Principal Actions
a. Estrogen and Progesterone	
b.	Increases flexibility of the pubic symphysis and helps to dilate cervix.
c.	Inhibits the secretion of FSH from the anterior pituitary.
d. Testosterone	

H. Pineal Gland (Epiphysis Cerebri)   (page 672)

H1.   Where is the pineal gland located?

H2.   One hormone secreted by the pineal gland is _____, which is liberated

during darkness and is thought to _____ sleep.

I. Thymus   (page 673)

I1.   List the four hormones produced by the thymus gland.

a.

b.

c.

d.

I2.   What is the general function of the thymus?

J. Other Endocrine Tissues   (page 673)

J1.   List the hormones synthesized by the gastrointestinal tract.

a.

b.

c.

d.

J2.   The kidneys release hormones called _____, and

_____.

J3.   The _____ produces human chorionic gonadotropin, estrogens,
progesterone, and human chorionic somatomammotropin.

J4.   The heart produces a hormone called ANP (_____

_____ _____), which functions to *(increase?*
decrease?) blood pressure.

K. Developmental Anatomy of the Endocrine System (pages 674–675)

K1. Match the endocrine gland with its embryologic origin.

AC	adrenal cortex	P	pancreas
AH	adenohypophysis	PN	pineal
AM	adrenal medulla	PT	parathyroid
NH	neurohypophysis	T	thyroid
	TH thymus		

a. _____ Arise from pharyngeal pouches (two answers).

b. _____ Derived from an outgrowth of ectoderm located on the floor of the hypothalamus.

c. _____ Arises from the floor of the pharynx.

d. _____ Arises from ectoderm of the diencephalon.

e. _____ Derived from intermediate mesoderm from the same region that produces the gonads.

f. _____ Derived from an outgrowth of the roof of the mouth called the hypophyseal pouch.

g. _____ Derives from the neural crest, which also produces sympathetic ganglia.

h. _____ Develops from the dorsal and ventral outgrowths of the foregut that later becomes the duodenum.

L. Aging and the Endocrine System (page 675)

L1. List some effects of aging on the endocrine system.

ANSWERS TO SELECT QUESTIONS

A1. Nervous, endocrine, hormones.
A3. Ducts, surface, capillaries (blood).
A4. Pituitary, thyroid, parathyroids, adrenals and pineal.
A5. Target, receptors.
A6. Decrease.
A7. (a) A—Pineal, B—parathyroid, C—adrenal, D—pituitary, E—thyroid, F—thymus, 1—skin, 2—liver, 3—pancreas, 4—small intestine, 5—heart, 6—stomach, 7—kidney, 8—ovaries, 9—testes.
B1. Master, posterior.
B2. Infundibulum.
B3. Adenohypophysis, neurohypophysis.
B4. Releasing hormones, inhibiting hormones.
B5. (a) SC; (b) GC; (c) TC; (d) LC; (e) CLC; (f) GC; (g) CLC.

B6. Tropic.
B7. Neurosecretory, antidiuretic, oxytocin.
B8. A—Posterior hypophyseal veins, B—posterior pituitary, C—inferior hypophyseal artery, D—superior hypophyseal artery, E—hypophyseal portal veins, F—anterior pituitary, G—anterior hypophyseal veins.
C1. Isthmus, 80, 120.
C2.

Cell	Hormone(s)
a. Follicular	Thyroxine (T4) Triiodothyronine (T3)
b. Parafollicular	Calcitonin

D1. Posterior, principal (chief), oxyphil.
D2. Increasing, Ca^{2+} and Mg^{2+}, decreasing, phosphate.
E1. Adrenal cortex, adrenal medulla.

E2.

Zone	Grouping
a. Glomerulosa	**Mineralocorticoids**
b. Fasciculata	**Glucocorticoids**
c. **Reticularis**	Gonadocorticoids

E3. Chromaffin, sympathetic.

E4. Epinephrine, norepinephrine.

F1. Endocrine, exocrine.

F2. Pancreatic islets; (a) F; (b) A; (c) D; (d) B; (e) F; (f) B; (g) A; (h) D.

F3. Acini.

F4. Secretin, cholecystokinin.

G1.

Hormone	Principal Actions
a. Estrogen and Progesterone	**Development and maintenance of feminine sexual characteristics. Regulates menstrual cycle and oogenesis, maintains pregnancy; prepares mammary glands for lactation.**
b. **Relaxin**	Increases flexibility of the pubic symphysis and helps to dilate cervix.
c. **Inhibin**	Inhibits the secretion of FSH from the anterior pituitary.
d. **Testosterone**	**Development and maintenance of masculine sexual characteristics. Regulates sperm production, stimulates descent of testes.**

H1. Attached to the roof of the third ventricle.

H2. Melatonin, promote.

I1. (a) Thymosin; (b) thymic humoral factor; (c) thymic factor; (d) thymopoietin.

I2. Promotes the proliferation and maturation of T cells, which destroy foreign microbes and substances.

J1. (a) Gastrin; (b) gastric inhibitory peptide; (c) secretin; (d) cholecystokinin.

J2. Erythropoietin, calcitriol.

J3. Placenta.

J4. Atrial natriuretic peptide, decrease.

K1. (a) PT, TH; (b) NH; (c) T; (d) PN; (e) AC; (f) AH; (g) AM; (h) P.

SELF QUIZ

Choose the one best answer to the following questions.

1. Which gland is responsible for the secretion of thyroid-stimulating hormone?

   A. thyroid
   B. hypothalamus
   C. anterior pituitary
   D. posterior pituitary
   E. A and D are correct

2. Exocrine glands release their secretions into a duct or onto a surface, whereas endocrine glands secrete into

   A. muscle tissue
   B. open cavities
   C. closed cavities
   D. extracellular space around the secretory cells
   E. none of the above are correct

3. Hormones produced by which gland promote the proliferation and maturation of T lymphocytes?

   A. anterior pituitary
   B. thyroid
   C. thymus
   D. adrenals
   E. posterior pituitary

4. Antidiuretic hormone and oxytocin are produced by the

   A. anterior pituitary
   B. posterior pituitary
   C. parathyroid glands
   D. hypothalamus
   E. A and D are correct

5. Giantism is associated with hypersecretion of a hormone from the

   A. anterior pituitary
   B. thyroid
   C. adrenals
   D. posterior pituitary
   E. testes

6. Thyroxine production is dependent upon adequate dietary intake of which ion?

   A. calcium
   B. potassium
   C. iodine
   D. magnesium
   E. sodium

7. If an individual has type I diabetes mellitus, which pancreatic cells are involved with this disorder?

   A. delta
   B. alpha
   C. beta
   D. F-cells
   E. none of the above are correct

8. Which of the following is a gonadotropin?

   A. ACTH
   B. FSH
   C. PRL
   D. MSH
   E. hGH

9. Addison's disease is associated with a disorder of the _____ gland.

   A. pituitary
   B. adrenal
   C. thyroid
   D. pancreas
   E. pineal

10. The principal hormone secreted by the cells of the zona glomerulosa is _____. (Hint: it is a mineralocorticoid.)

    A. cortisol
    B. aldosterone
    C. estrogen
    D. androgens
    E. C and D are both correct

Answer (T) True or (F) False to the following questions.

11. _____ Hormones are released by target cells and are carried by the blood to secretory cells.

12. _____ The follicular cells of the thyroid gland are responsible for the secretion of calcitonin.

13. _____ Myxedema is associated with hyperthyroidism during the adult years.

14. _____ The secretion of parathyroid hormone (PTH) increases the blood calcium levels.

15. _____ The release of hormones from the adrenal medulla would produce effects similar to stimulation by the sympathetic division of the ANS.

16. _____ Secretion of glucagon leads to a hypoglycemic state.

17. _____ A principal effect of oxytocin is to stimulate uterine contraction during childbirth.

18. _____ The testes function only as an endocrine gland.

19. _____ Diabetes insipidus is a result of hypersecretion of antidiuretic hormone (ADH).

Fill in the blanks.

20. _____, released from the neurohypophysis, initiates milk production by the mammary glands.

21. Hormones that increase the rate of protein catabolism, stimulate gluconeogenesis and lipolysis, provide

   resistance to stress, and dampen inflammation are produced in the zona _____.

22. _____ cells of the pancreas secrete a hormone which inhibits the secretion of insulin and glucagon.

23. A _____ is simply an enlarged thyroid gland.

24. _____ disease, an autoimmune disorder, is the most common form of hyperthyroidism.

25. Melatonin is secreted by the _____ gland.

ANSWERS TO THE SELF QUIZ

1. C
2. D
3. C
4. D
5. A
6. C
7. C
8. B
9. B

10. B
11. F
12. F
13. F
14. T
15. T
16. F
17. T
18. F

19. F
20. Prolactin (PRL)
21. Fasciculata
22. Delta
23. Goiter
24. Graves'
25. Pineal

The Respiratory System

CHAPTER 23

SYNOPSIS

On your journey through the human body it should have become apparent that, despite the complex nature of each system, no one system can survive without the others.

Every cell in our body requires oxygen (O_2); it is essential for cellular metabolism and the release of energy. The metabolic reactions which occur in the cells also lead to the release of carbon dioxide (CO_2) as a by-product. The **cardiovascular system's** function is to deliver oxygen to and remove carbon dioxide from the cells. The exchange of oxygen and carbon dioxide between our body and the environment occurs via the **respiratory system.** It is the highly coordinated action of these two systems that provides the precise level of blood gases necessary for optimal body function.

Although the exchange of gases is undoubtedly the respiratory system's primary function, it also contains receptors for olfaction (sense of smell), filters inspired air, enables vocalization, and helps eliminate waste.

The **nose**, **pharynx**, **larynx**, **trachea**, **bronchi**, **bronchioles**, and **terminal bronchioles** serve as the conducting portion of the respiratory system, carrying gases to and from the lungs. The respiratory portion (the site of gas exchange) consists of segments of the lungs, including the **respiratory bronchioles**, **alveolar ducts**, **alveolar sacs**, and an estimated 300 million **alveoli**, providing a surface area of 70 m² for the exchange of gases.

This wondrously designed system is not without its share of difficulties. The respiratory system, due to its constant exposure to the environment, is susceptible to a variety of microbial, viral, and chemical disorders (such as bronchitis, pneumonia, emphysema, and tuberculosis).

TOPIC OUTLINE AND OBJECTIVES

A. Introduction and Organs of Respiration

1. Identify the organs of the respiratory system.
2. Contrast the internal and external structures of the nose.
3. List and describe the three anatomical regions of the pharynx.
4. Describe the structural features of the larynx and their correlation to respiratory function and vocalization.
5. Describe the location and structural features of the trachea and bronchi.
6. List and describe the coverings of the lung.
7. Examine the gross anatomy and divisions of the lung and the components of the lobules.
8. Describe the alveolar–capillary membrane and explain its function in gas exchange.

B. Mechanics of Pulmonary Ventilation (Breathing)

C. Control of Respiration

D. Developmental Anatomy of the Respiratory System

E. Aging and the Respiratory System

F. Applications to Health

9. Define asthma, bronchitis, emphysema, lung cancer, pneumonia, tuberculosis, respiratory distress syndrome (RDS), respiratory failure, coryza and influenza, pulmonary embolism, and cystic fibrosis.

G. Key Medical Terms Associated with the Respiratory System

SCIENTIFIC TERMINOLOGY

Find an anatomical sample word for each prefix and suffix:

Prefix/Suffix	Meaning	Sample Word
arytaina-	ladle	
ateles-	incomplete	
bronchus-	windpipe	
dys-	painful	
glotta-	tongue	
hypo-	below	
ortho-	straight	
pneumo-	air of breath	
pulmo-	lung	
rhino-	nose	
tachy-	rapid	

A. Introduction and Organs of Respiration (pages 681–696)

A1. In addition to its vital role in oxygen and carbon dioxide exchange, the respiratory system has several other functions. List these functions below.

a.

b.

c.

d.

A2. The upper respiratory system consists of the _____,

_____, and associated structures. The lower respiratory system refers

to the _____, _____, _____, and

lungs.

A3. The _____ portion of the respiratory system includes the interconnecting cavities and tubes that conduct air into the lungs. The

_____ portion is the site of respiratory gas exchange.

A4. The external portion of the nose consists of a supporting framework of

_____ and _____ cartilage covered with muscle and skin.

A5. Answer the following questions about the nose.

a. The internal portion of the nose communicates posteriorly with the pharynx via two

openings called the _____ _____.

b. List the three functions of the interior structures of the nose.

1.

2.

3.

c. The olfactory epithelium of the nasal cavity is made of a *(stratified squamous? pseudo-stratified columnar?)* epithelium. Extending into the nasal cavity are the superior, middle, and inferior nasal conchae, which form groovelike passageways called the

_____.

d. The receptors of olfactory sensation are found in the *(walls? roof? floor?)* of the nasal cavity.

A6. Based upon the anatomy of the pharynx, offer an explanation of why swallowing, when ascending in a jet, would cause your ears to "pop"?

A7. Match the tonsil to the proper location. (See Figure 23.2 in the textbook for some help.)

LG lingual tonsils	PL palatine tonsils	PH pharyngeal tonsil

a. _____ Located in the posterior wall of the nasopharynx.

b. _____ Lies in the lateral walls of the oropharynx.

c. _____ Located at the base of the tongue.

A8. The larynx or _____ _____ lies in the midline of the neck anterior to the

(C3–C5? C4–C6?) vertebrae. Its wall is composed of _____ pieces of cartilage.

A9. Explain how the larynx prevents food from entering the trachea.

A10. Of the paired cartilages of the larynx, arytenoid cartilages are considered the most important. Why?

A11. The upper pair of mucous membrane folds in the larynx are called the

_____ folds, and the lower pair are referred to as the

_____ folds. The space between the upper folds is known as the

_____ _____.

A12. Briefly explain how the larynx produces sound. In your explanation include how pitch is controlled and what causes the difference in pitch between males and females.

A13. Refer to Figure LG 23.1 and identify the following structures.

_____ conchae	_____ opening of auditory tube
_____ cricoid cartilage	_____ oral cavity
_____ epiglottis	_____ oropharynx
_____ esophagus	_____ palatine tonsil
_____ external naris	_____ paranasal sinuses
_____ hard palate	_____ pharyngeal tonsil
_____ hyoid bone	_____ soft palate
_____ internal naris	_____ thyroid cartilage
_____ laryngopharynx	_____ tongue
_____ lingual tonsil	_____ trachea
_____ nasopharynx	_____ true vocal cords

A14. Answer these questions pertaining to the trachea.

a. The trachea is located *(anterior? posterior?)* to the esophagus and is commonly referred to

as the _____. It extends from the larynx to the *(5th? 6th?)* thoracic

vertebra, where it divides into the two primary _____.

b. There are _____ to _____ C-shaped cartilage rings that support the trachea.

Transverse smooth muscle fibers, called the _____ muscle, attach to the open ends of these cartilage rings.

c. The internal ridge at the point where the trachea divides into the primary bronchi is called

the _____. A widening and distortion of this structure is often indicative

of what type of disorder? _____

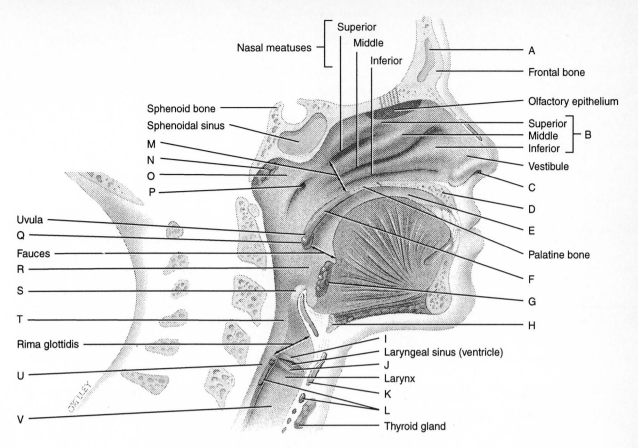

Diagram of sagittal section of the left side of the head
and neck showing the location of respiratory structures

Figure LG 23.1 Sagittal section of the left side of head and neck.

A15. Explain why the entry of a foreign object into the lungs is more likely to occur through
the right primary bronchus rather than the left primary bronchus.

A16. As the branching of the bronchial tree continues, one would expect to see an *(increase?
decrease?)* in cartilaginous structures, while the amount of smooth muscle

_____. Why is this structural variation of importance to an asthmatic?

A17. Answer the following questions about the lungs.

a. Located between the parietal and visceral pleura is the pleural cavity, which contains a

   _____ _____.

b. The inferior portion of the lung, which rests on the diaphragm, is the

   _____, whereas the narrow superior portion is called the

   _____.

c. Primary bronchi, pulmonary blood vessels, lymphatic vessels, and nerves enter and exit the

   lungs through the _____.

A18. Using the terms *left* (L) or *right* (R), answer these questions about the lungs.

a. _____ has a cardiac notch   c. _____ has two lobes

b. _____ has a horizontal fissure   d. _____ has three lobes

A19. Refer to Figure LG 23.2 and identify the following structures.

_____ bronchiole   _____ primary bronchus

_____ carina   _____ secondary bronchus

_____ diaphragm   _____ terminal bronchiole

_____ larynx   _____ tertiary bronchi

_____ parietal pleura   _____ trachea

_____ pleural cavity   _____ visceral pleura

A20. Check your understanding of the lobules by answering the following questions.

a. Place the following structures of the lobule in the correct order.

A	alveoli	AD	alveolar duct
RB	respiratory bronchioles	TB	terminal bronchiole

   _____ → _____ → _____ → _____

b. Type II alveolar cells produce a phospholipid and lipoprotein substance called

   _____. What is this substance's function?

c. In order for the diffusion of oxygen and carbon dioxide to occur, the respiratory gases must cross the alveolar–capillary (respiratory) membrane. Identify the four structures that form this membrane.

   1.   3.

   2.   4.

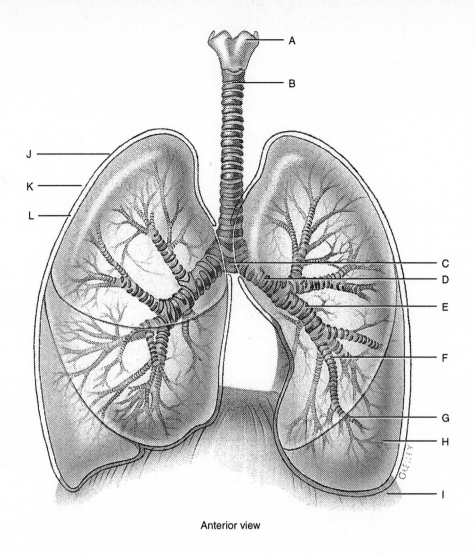

Anterior view

Figure LG 23.2 Diagram of lungs with pleural coverings and bronchial tree.

A21. The lungs possess a double blood supply, with deoxygenated blood passing through the

_____ trunk and pulmonary arteries and oxygenated blood entering

the lungs via the _____ arteries.

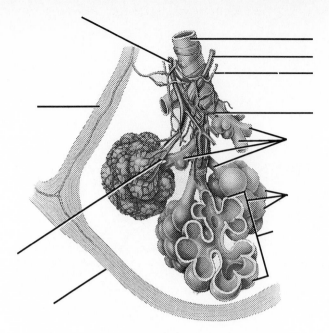

Portion of a lobule of the lung

Figure LG 23.3 Histology of the lungs.

A22. Refer to Figure LG 23.3 and label the following structures using the leader lines.

alveolar ducts   pulmonary capillaries
alveolar sacs   pulmonary venule
alveoli   respiratory bronchiole
elastic connective tissue   terminal bronchiole
lymphatic vessels   visceral pleura
pulmonary arteriole

B. Mechanics of Pulmonary Ventilation (Breathing) (pages 697–698)

B1. List and define the three basic processes involved in respiration.

a.

b.

c.

B2. Just prior to each inspiration, the air pressure inside the lungs *(is greater than? equals? is less than?)* the pressure of the atmosphere.

B3. For air to flow into the lungs, pressure in the alveoli must become *(lower? higher?)* than the pressure in the atmosphere.

B4. The _____ is the most important muscle of inspiration. Contraction of this muscle causes it to flatten, leading to a(n) *(decrease? increase?)* in the vertical dimension of the thoracic cavity.

B5. Deep, labored inspiration requires the participation of several accessory muscles,

including the sternocleidomastoid, _____ and _____ minor.

B6. Normal expiration is dependent upon two factors. Describe these factors below.

a.

b.

C. Control of Respiration   (pages 698–701)

C1. The area located bilaterally in the medulla oblongata and pons from which nerve

impulses are sent to the respiratory muscles is called the _____

_____.

C2. The area described in question C1 is functionally divided into three areas. Name these areas below.

a.

b.

c.

C3. Match the respiratory area with the correct description.

> AA   apneustic area
> MRA   medullary rhythmicity area
> PA   pneumotaxic area

a. _____ This area stimulates the inspiratory area to prolong inspiration, thus inhibiting expiration.

b. _____ The basic rhythm of respiration is determined by nerve impulses generated here.

c. _____ This area transmits inhibitory impulses to the inspiratory area, helping to turn off the inspiratory area before the lungs become too full of air.

C4. The chemosensitive area within the medulla oblongata is composed of a group of neurons that are highly sensitive to blood concentrations of *(O_2? CO_2?).*

C5. Discuss the function of the stretch receptors and their relationship to the inflation reflex.

D. Developmental Anatomy of the Respiratory System   (page 702)

D1. About the *(3rd? 4th?)* week of fetal development, the respiratory system begins as an outgrowth of the *(mesoderm? endoderm?)* of the foregut just behind the pharynx.

D2. The cartilage, connective tissue, and smooth muscle of the bronchial tubes and pleural

sacs are contributed by the _____ cells.

E. Aging and the Respiratory System (page 702)

E1. With age the airways, air sacs, and chest wall become *(more? less?)* elastic, resulting in a(n) *(increase? decrease?)* in pulmonary lung capacity.

F. Applications to Health (Refer to the *Application to Health* companion to answer Part F questions.)

F1. Match the respiratory disorder with the correct description.

A asthma	P pneumonia
BR bronchitis	TB tuberculosis
E emphysema	

a. _____ A communicable disease that most often affects the lungs and the pleurae. Bacteria destroy parts of the lung tissue and replace it with fibrous connective tissue.

b. _____ An inflammatory disorder characterized by enlargement of glands and goblet cells lining the bronchial airways. The typical symptom is a productive cough.

c. _____ The alveolar walls disintegrate, producing abnormally large air spaces that remain filled with air during expiration.

d. _____ Possibly an allergic reaction, characterized by attacks of wheezing and difficult breathing, with bronchospasm and increased mucus secretion.

e. _____ An acute infection or inflammation of the alveoli in which the sacs fill up with fluid and dead white blood cells. Edema is a cardinal sign of this disorder.

F2. This disorder affects 65,000 premature babies born in the U.S. every year. Prior to birth, the respiratory passages fill with fluid. Name this disorder.

_____ _____ _____ .

F3. List four causes of respiratory failure.

a.

b.

c.

d.

F4. This disorder is caused by a group of viruses called rhinoviruses. It is characterized by sneezing, excessive nasal secretion, dry cough, and congestion. Name this disorder.

_____

F5. A blood clot or other foreign substance that obstructs a pulmonary arterial vessel is called

a _____ _____ .

F6. The prime agent in the development of bronchogenic carcinoma is

_____ . How does this affect the basal cells and mucus production?

ANSWERS TO SELECT QUESTIONS

A1. (a) Sense of smell; (b) filters air; (c) produces sound; (d) eliminates waste.

A2. Nose, pharynx; larynx, trachea, bronchi.

A3. Conducting, respiratory.

A4. Bone, hyaline.

A5. (a) Internal nares; (b) 1. warm, moisten, and filter incoming air, 2. receive olfactory stimuli, 3. modify speech sounds; (c) pseudostratified columnar, meatuses; (d) roof.

A7. (a) PH; (b) PL; (c) LG.

A8. Voice box, C4–C6, nine.

A10. They influence the positions and tensions of the vocal folds.

A11. Ventricular, vocal, rima vestibuli.

A13. A—Frontal sinus, B—conchae, C—external naris, D—maxilla, E—oral cavity, F—soft palate, G—lingual tonsil, H—hyoid bone, I—ventricular fold, J—true vocal cord, K—thyroid cartilage, L—cricoid cartilage, M—internal naris, N—pharyngeal tonsil, O—nasopharynx, P—opening of auditory tube, Q—palatine tonsil, R—oropharynx, S—epiglottis, T—laryngopharynx, U—esophagus, V—trachea.

A14. (a) Anterior, windpipe, 5th, bronchi; (b) 16, 20, trachealis; (c) carina, carcinoma of the lymph nodes around trachea.

A15. Right primary bronchi is more vertical, shorter, and wider than the left.

A16. Decrease, increases; Asthma is brought on by spasms of the smooth muscle in the walls of the smaller bronchi and bronchioles, causing the passageways to close partially or completely.

A17. (a) Lubricating (serous) fluid; (b) base, apex; (c) hilus.

A18. (a) L; (b) R; (c) L; (d) R.

A19. A—Larynx, B—trachea, C—carina, D—primary bronchus, E—secondary bronchus, F—tertiary bronchus, G—bronchiole, H—terminal bronchiole, I—diaphragm, J—visceral pleura, K—parietal pleura, L—pleural cavity.

A20. (a) TB > RB > AD > A; (b) surfactant; (c) 1. alveolar epithelial wall, 2. epithelial basement membrane, 3. capillary basement membrane, 4. endothelial cells.

A21. Pulmonary, bronchial.

B1. (a) Pulmonary ventilation, (b) external respiration, (c) internal respiration.

B2. Equals.

B3. Lower.

B4. Diaphragm, increase.

B5. Scalenes, pectoralis.

B6. (a) The recoil of elastic fibers that were stretched during inspiration; (b) the inward pull of surface tension due to the film of alveolar fluid.

C1. Respiratory center.

C2. (a) Medullary rhythmicity area in the medulla oblongata; (b) pneumotaxic area in the pons; (c) apneustic area in the pons.

C3. (a) AA; (b) MRA; (c) PA.

C4. CO_2.

D1. 4th, endoderm.

D2. Mesenchymal (mesodermal).

E1. Less, decrease.

F1. (a) TB; (b) BR; (c) E; (d) A; (e) P.

F2. Respiratory distress syndrome (RDS).

F3. 1. Lung disorders, 2. mechanical disorders that disturb the chest wall, 3. depression of respiratory center by drugs, strokes, or trauma, 4. carbon monoxide poisoning.

F4. Coryza (common cold).

F5. Pulmonary embolism.

F6. Smoking (pollutants).

SELF QUIZ

Choose the one best answer to the following questions.

1. Deoxygenated blood becomes oxygenated in which portion of the bronchial tree?

   A. primary bronchi   C. trachea
   B. secondary bronchi   D. bronchioles
       E. alveoli

2. The auditory (Eustachian) tube connects the middle ear with the _____.

   A. nasal cavity   D. laryngopharynx
   B. nasopharynx   E. larynx
   C. oropharynx

3. The blood with the highest oxygen content is found in the

A. superior vena cava
B. pulmonary veins
C. pulmonary arteries
D. inferior vena cava
E. coronary sinus

4. The vocal folds are attached to which cartilage structures?

A. cricoid
B. corniculate
C. arytenoid
D. cuneiform
E. thyroid

5. The exchange of oxygen and carbon dioxide between the lungs and the pulmonary blood occurs by the process of

A. osmosis
B. facilitated diffusion
C. diffusion
D. active transport
E. phagocytosis

6. Which structure does NOT belong with the others?

A. nose
B. larynx
C. trachea
D. bronchi
E. lungs

7. This structure is commonly called the "Adam's apple."

A. larynx
B. epiglottis
C. thyroid cartilage
D. cricoid cartilage
E. arytenoid cartilage

8. A 65-year-old male comes to your office complaining of difficulty breathing. He indicates that he has smoked cigarettes for over 45 years. Your examination indicates an enlargement of the heart, barrel chest, and decreased pulmonary capacity. What is your diagnosis?

A. asthma
B. emphysema
C. bronchitis
D. pneumonia
E. coryza

9. The olfactory area is located

A. within the superior nasal conchae
B. below the superior nasal conchae
C. in the middle nasal meatus
D. in the inferior nasal meatus
E. C and D are both correct

10. The last structure(s) that oxygen (in the lungs) must cross to enter the cardiovascular system is (are) the

A. type I alveolar cells
B. type II alveolar cells
C. pulmonary epithelial basement membrane
D. capillary basement membrane
E. endothelial cells

11. Pneumothorax refers to the pleural cavity being filled with

A. serous fluid
B. blood
C. air
D. water
E. mucus

Arrange the answers in the correct order.

12. From superficial to deepest.   _____ _____ _____
    A. pleural cavity
    B. visceral pleura
    C. parietal pleura

13. From superior to inferior.   _____ _____ _____ _____ _____
    A. bronchioles
    B. trachea
    C. primary bronchi
    D. terminal bronchiole
    E. alveolar ducts

Answer (T) True or (F) False to the following questions.

14. _____ The vocal folds are located in the larynx.

15. _____ Type I alveolar cells secrete a phospholipid substance called a surfactant.

16. _____ Nebulization is a procedure that administers medication to the respiratory tract in the form of droplets suspended in air.

17. _____ The right lung possesses both a horizontal and an oblique fissure.

18. _____ There are three primary bronchi associated with the left lung.

19. _____ A tracheostomy is made superior to the cricoid cartilage.

Fill in the blanks.

20. Inflammation of the pleural membrane is commonly called _____.

21. The _____ aids in routing liquids and foods into the esophagus, keeping them out of the larynx.

22. _____ refers to a reduction in oxygen supply to cells.

23. The inhalation of a foreign substance such as water, food, or a foreign body into the bronchial tree is called

_____.

24. There are ten _____ _____ in each lung.

25. A commonly performed cosmetic surgery that alters the external nasal structure is called

_____.

ANSWERS TO THE SELF QUIZ

1. E
2. B
3. B
4. C
5. C
6. A
7. C
8. B
9. A

10. E
11. C
12. C, A, B
13. B, C, A, D, E
14. T
15. F
16. T
17. T
18. F

19. F
20. Pleurisy
21. Epiglottis
22. Hypoxia
23. Aspiration
24. Tertiary bronchi or bronchopul-
    monary segments
25. Rhinoplasty

The Digestive System

CHAPTER

24

SYNOPSIS

The interdependency of systems is nowhere more apparent than with the digestive system. Food is essential for providing the materials necessary for the repair, replacement, and growth of tissue and as the source of energy that fuels the metabolic reactions in every cell. The answer to the question of whether we "eat to live" or "live to eat" is the former.

    The process of digestion involves the conversion of large food substances into particles small enough to enter the bloodstream and cells. This process occurs through a series of complex mechanical and chemical mechanisms. The **gastrointestinal (GI) tract**, which runs from the oral cavity to the anus, provides the environment in which our ingested foods are converted and absorbed into the body. Working in concert with the gastrointestinal tract are the accessory structures: the **tongue**, **teeth**, **salivary glands**, **pancreas**, **liver**, and **gallbladder**. The tongue and teeth assist in chewing and swallowing, while the other accessory structures are responsible for the production and storage of secretions that aid in the chemical breakdown of food.

    In addition to the aforementioned functions, the digestive system serves as a nonspecific defense mechanism, resisting mechanical stress and the ingestion of pathogens along with swallowed food.

TOPIC OUTLINE AND OBJECTIVES

A. Organization

    1. List the components of the gastrointestinal tract and the accessory structures.

B. Digestive Processes

    2. List the six basic digestive processes.
    3. Differentiate between chemical and mechanical digestion.

C. General Histology of the GI Tract

    4. Describe the general histology of the four layers of the gastrointestinal tract.
    5. Discuss the components of the peritoneum.

D. Mouth (Oral Cavity)

    6. Describe the boundaries and structures associated with the oral cavity.

    7. Describe the structure and function of the tongue.
    8. Identify the location and histology of the salivary glands.
    9. Identify the parts of a typical tooth, and compare the primary and secondary dentitions.

E. Pharynx and Esophagus

    10. Describe the histology of the pharynx.
    11. Describe the histology, function, and blood and nerve supply of the esophagus.

F. Stomach

    12. Describe the structure and functions of the stomach in digestion.

G. Pancreas, Liver, and Gallbladder—The Accessory Structures

13. Describe the histology, function, and blood and nerve supply of the pancreas, liver, and gallbladder.

H. Small Intestine

14. Describe the structure of the small intestine and how it is modified for digestion and absorption.

I. Large Intestine

15. Describe the histology of the large intestine and discuss its function in digestion and defecation.

J. Developmental Anatomy of the Digestive System

K. Aging and the Digestive System

L. Applications to Health

16. Describe the etiology and/or the symptomatology associated with dental caries, periodontal disease, peptic ulcer disease, colorectal cancer, cirrhosis, hepatitis, gallstones, obesity, anorexia nervosa, bulimia, and diarrhea.

M. Key Medical Terms Associated with the Digestive System

SCIENTIFIC TERMINOLOGY

Find an anatomical sample word for each prefix and suffix:

Prefix/Suffix	Meaning	Sample Word
-chalsis	relaxation	
chole-	bile	
enteron-	intestines	
galla-	bile	
gastro-	stomach	
ortho-	straight	
-otid	ear	
peri-	around	
retro-	backward or located behind	
-stalsis	contraction	

A. Organization   (page 708)

A1. Complete the list below of the organs of the gastrointestinal tract and accessory structures.

Gastrointestinal Tract

a. _____

b. _____

c. _____

d. _____

e. _____

f. _____

Accessory Structures

g. _____

h. _____

i. _____

j. _____

k. _____

l. _____

B. Digestive Processes   (page 708)

B1.   Match the digestive process with the correct definition

A	absorption	I	ingestion
D	digestion	P	propulsion
DF	defecation	S	secretion

a. _____ Elimination of indigestible substances.

b. _____ Passage of food along the gastrointestinal tract toward the anus.

c. _____ Breakdown of food by chemical and mechanical processes.

d. _____ Taking food into the mouth.

e. _____ Passage of digested food into the cardiovascular or lymphatic system.

f. _____ Production and release of about 9 liters of water, acid, buffers, and enzymes per day.

B2.   Define the following terms.

a. Chemical digestion

b. Mechanical digestion

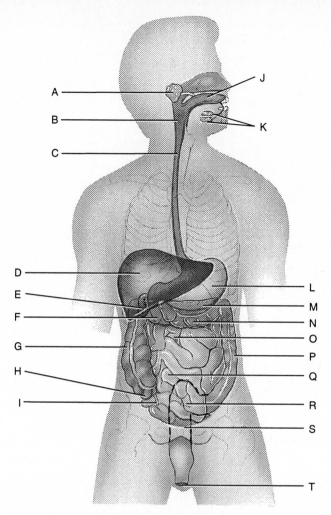

Right lateral view of head and neck and anterior view of chest, abdomen, and pelvis showing organs of the digestive system

Figure LG 24.1   Organs of the digestive system.

B3.   Refer to Figure LG 24.1 and match the following structures with the corresponding letter.

_____ anus

_____ cecum

_____ colon, ascending

_____ colon, descending

_____ colon, sigmoid

_____ colon, transverse

_____ duodenum

_____ esophagus

_____ gallbladder

_____ ileum

_____ jejunum

_____ liver

_____ mouth

_____ pancreas

_____ parotid gland

_____ pharynx

_____ rectum

_____ stomach

_____ sublingual and submandibular glands

_____ vermiform appendix

C. General Histology of the GI Tract   (page 708–713)

C1.   Answer these questions about the general histology of the gastrointestinal tract.

a. The _____ is the inner lining of the tract. It is composed of *(2? 3?)* layers.

b. The layer of the mucosa that contains many blood and lymph vessels and scattered

   lymphatic nodules is the _____ _____.

c. With the exception of the mouth, esophagus, and anal canal, the epithelial layer of the

   mucosa is *(stratified? simple?)*. This allows for the processes of _____

   and _____ to occur.

d. Name the autonomic nerve supply, found in the submucosa, that innervates the

   muscularis mucosae. _____ _____

e. The muscularis from the lower esophagus through the large intestine is generally found in

   two sheets: an inner sheet of _____ fibers and an outer sheet of

   _____ fibers.

f. The outermost layer of the GI tract below the diaphragm is called the serosa or

   _____ _____.

g. The accumulation of serous fluids in the peritoneal cavity is called *(edema? ascites?)*

h. What is meant by the term "retroperitoneal"?

C2.   Refer to Figure LG 24.2 and answer the following questions.

a. Identify the following structures, labeled A–G.

   _____ circular muscle   _____ longitudinal muscle

   _____ epithelium   _____ muscularis mucosae

   _____ gland in submucosa   _____ villus

   _____ lamina propria

b. What layers are represented by numbers 1–4 in this figure?

   1.

   2.

   3.

   4.

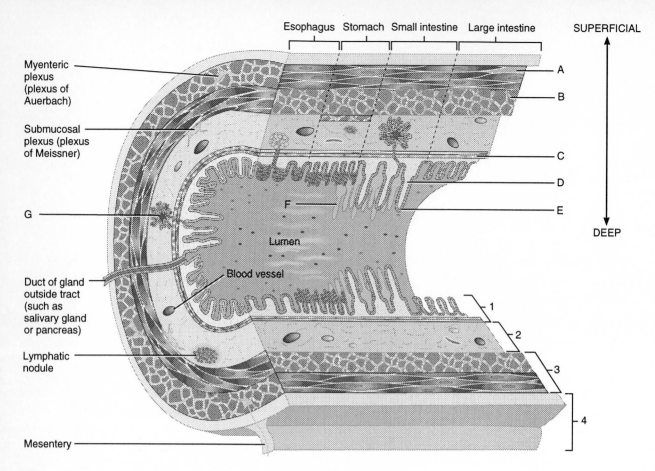

Figure LG 24.2   Composite of various sections of the gastrointestinal tract.

C3.   Match the peritoneal extension with the proper description.

> FL   falciform ligament
> GO   greater omentum
> LO   lesser omentum
> MC   mesocolon
> MS   mesentery

a.   _____ Binds the large intestine to the posterior abdominal wall.

b.   _____ Attaches the liver to the anterior abdominal wall.

c.   _____ Drapes over the transverse colon and coils of the small intestine.

d.   _____ Binds the small intestine to the posterior abdominal wall.

e.   _____ Suspends the stomach and duodenum from the liver.

D. Mouth (Oral Cavity)   (pages 713–719)

D1. Match the following terms about the oral cavity with the definitions.

F	fauces		U	uvula
LB	labia		V	vermilion
LF	labial frenulum		VS	vestibule

a. _____ Conical muscular process of the soft palate.

b. _____ Fleshy folds surrounding the opening of the mouth.

c. _____ Space bound externally by the cheeks and lips, and internally by the gums and teeth.

d. _____ The opening between the oral cavity and the pharynx.

e. _____ Attaches the inner surface of each lip to the gum.

f. _____ Transition zone where the outer skin of the lips meets the inner zone (mucous membrane).

D2. The palatine tonsils are situated between the _____ and

_____ arches.

D3. The skeletal muscle of the tongue is divided into (E) extrinsic and (I) intrinsic muscles. Identify the movements controlled by these muscles.

a. _____ Alters shape and size of the tongue for speech and swallowing.

b. _____ Moves the tongue from side to side and in and out.

D4. List the three types of papillae found on the upper surface and sides of the tongue. Which one does not normally have taste buds?

a.

b.

c.

D5. Complete this exercise about the salivary glands.

a. The glands responsible for the secretion of saliva include the _____ glands, which are small glands that line the cheeks, and the three major pairs of salivary

glands: _____, _____, and _____ glands.

b. State the functions of saliva.

c. The salivary glands receive both sympathetic and parasympathetic innervation. The *(sympathetic? parasympathetic?)* fibers form plexuses on the blood vessels that supply the glands

and initiate _____.

d. Name the enzyme present in saliva that initiates the breakdown of starch:

_____ _____

e. Identify the three salivary glands on Figure LG 24.1.

D6. What is the function of the periodontal ligament?

D7. Match the parts of the tooth with the correct description.

AF	apical foramen	N	neck
C	crown	P	pulp
CM	cementum	R	root
D	dentin	RC	root canal
E	enamel		

a. _____ A bonelike substance that attaches the root to the periodontal ligament.

b. _____ The exposed portion above the level of the gums.

c. _____ The hardest substance in the body; protects the tooth from wear.

d. _____ Consists of one to three projections embedded in the socket.

e. _____ Opening in the base of the root canal.

f. _____ Constricted junction line between the crown and root.

g. _____ Narrow extensions of the pulp cavity running through the root of the tooth.

h. _____ A calcified connective tissue that gives the tooth its basic shape and rigidity.

i. _____ A connective tissue containing blood vessels, nerves, and lymphatic vessels.

D8. Sensory fibers that innervate the teeth are branches of the *(V? VII?)* cranial nerve.

D9. Deciduous teeth begin to erupt at approximately the _____ month until all _____

teeth are present. They will eventually be replaced by _____ permanent teeth.

E. Pharynx and Esophagus (pages 719–721)

E1. Chewing or _____ results in the formation of a flexible mass of food

called a _____.

E2. The *(laryngo? naso? oro?)* pharynx has a respiratory function only.

E3. Swallowing or _____ moves the food from the mouth to the stomach.

E4. The esophagus is located *(anterior? posterior?)* to the trachea. It pierces the diaphragm

through an opening called the _____ _____.

E5. Briefly explain the function of the esophagus.

E6. _____ is defined as the involuntary muscular movements which propel the food through the digestive system.

E7. Failure of this structure to relax during swallowing would prevent the bolus from entering

the stomach. _____ _____ _____

E8. Complete the table below, indicating the mechanical and chemical digestion that takes place in the mouth, pharynx, and esophagus.

F. Stomach   (pages 721–724)

	Mechanical	**Chemical**
Mouth		Initiate starch digestion
Pharynx		
Esophagus	Peristalsis	

F1. List the correct order of food passage through the stomach.

body	fundus
cardia	pylorus

_____ → _____ → _____ → _____

F2. The structure of two layers of the stomach wall differ from that of the other parts of the GI tract. How do these layers differ?

a. Mucosa

b. Muscularis

F3. The gastric pits of the stomach are lined with four types of secreting cells. Complete the table below pertaining to these cells.

Cell Name	**Function**	**Secretion**
a. Chief (zymogenic)		
b.	Lubricates and protects	
c.		Hydrochloric acid and intrinsic factor
d. Enteroendocrine (G-cell)		

F4. In the stomach, the peristaltic movements are called _____

_____. This action along with the gastric secretions reduces the food to

a thin liquid called _____.

F5. Answer these questions about chemical digestion in the stomach.

a. The principal chemical activity of the stomach is to initiate the digestion of

_____. Another enzyme, gastric lipase, has a(n) *(limited? unlimited?)* role in the digestion of fats.

b. The contents of the stomach empty into the duodenum in _____ to _____ hours.

Foods richest in _____ leave first, followed by _____-

rich foods, and then foods high in _____.

G. Pancreas, Liver, and Gallbladder—The Accessory Structures   (pages 724–729)

G1. Complete these questions about the pancreas.

a. The pancreas is located *(anterior? posterior?)* to the *(lesser? greater?)* curvature of the

stomach. It is divided into three portions, with the _____ being closest to the C-shaped curve of the duodenum.

b. Two ducts convey the pancreatic secretions into the small intestine. Which duct unites with

the common bile duct at the hepatopancreatic ampulla? _____

_____

c. _____ _____ form the endocrine portion of the pancreas and consist of four types of cells. Match the secretions with the cells. (See Chapter 22.)

G glucagon	PP pancreatic polypeptide
I insulin	SS somatostatin

_____ alpha cells   _____ delta cells
_____ beta cells   _____ F-cells

d. Pancreatic juice is *(acidic? alkaline?)* since it contains _____ bicarbonate.

e. The pancreatic enzyme responsible for the digestion of carbohydrates is

_____ _____, while trypsin, chymotrypsin, and

carboxypeptidase digest _____.

G2. Answer these questions about the liver.

a. The liver occupies most of the right _____ and part of the

_____ regions of the abdominopelvic cavity.

b. The free end of the falciform ligament is the _____

_____, which extends from the liver to the _____.

c. The liver is made up of functional units called lobules. Each lobule consists of cells called

_____, which are arranged in branching plates around a

_____ vein. Between these irregular branching plates are spaces called

_____, which are partly lined with phagocytic cells called

_____ _____ cells.

G3. Using the abbreviations below, indicate the proper course bile travels from the hepatocytes to the hepatopancreatic ampulla.

BC	bile canaliculi	CBD	common bile duct
CHD	common hepatic duct	CD	cystic duct
RLHD	right and left hepatic ducts		

_____ → _____ → _____ joins with _____ to form _____

G4. The liver plays important roles in carbohydrate, fat, and protein metabolism. Briefly explain the liver's vital role in the metabolism of proteins.

G5. Name the vitamins and minerals that are stored in the liver.

G6. The liver has a dual blood supply. The vessel that supplies oxygenated blood is the

_____ _____, whereas the _____

_____ _____ carries deoxygenated blood containing newly absorbed nutrients from the GI tract.

G7. Answer these questions about the gallbladder.

a. Upon histological examination you would expect to find the (*mucosa? submucosa? muscularis?*) absent.

b. The gallbladder stores and concentrates bile (up to _____-fold) until it is needed in the small intestine.

c. The nerves to the gallbladder include branches from the _____ plexus

and CN (X) _____ nerve.

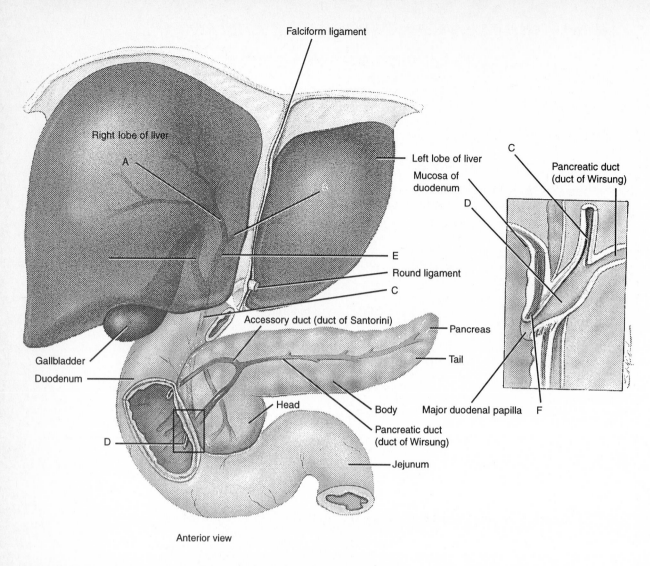

Falciform ligament

Right lobe of liver

A

Left lobe of liver

Mucosa of
duodenum

D

E

Round ligament

C

Accessory duct (duct of Santorini)

Pancreas

Tail

Gallbladder

Duodenum

Head

Body

Major duodenal papilla

F

Pancreatic duct
(duct of Wirsung)

Jejunum

C

Pancreatic duct
(duct of Wirsung)

D

Anterior view

Figure LG 24.3   Relation of pancreas to liver, gallbladder, and duodenum.

G8.   Refer to Figure LG 24.3 and identify the following structures.

_____ common bile duct   _____ left hepatic duct

_____ common hepatic duct   _____ right hepatic duct

_____ hepatopancreatic ampulla   _____ sphincter of the hepatopancreatic ampulla

H. Small Intestine   (pages 730–732)

H1.   Complete these questions about the anatomy and histology of the small intestine.

a.   The final portion of the small intestine is the _____ and the shortest part

is the _____. At death, the _____ is found empty.

b.   Special features of both the _____ and _____ allow
for the completion of the digestive and absorptive processes.

c. The intestinal glands are responsible for the secretion of _____

_____, while the duodenal glands secrete an _____
mucus that protects the wall of the small intestine.

d. Two vital structural variations in the mucosa of the small intestine that provide for an

enormous increase in the surface area are the _____ and

_____ (brush border).

e. Embedded in the connective tissue of a villus are an arteriole, a venule, a

_____ network, and a _____, which is a lymphatic
vessel.

H2. Name two types of lymphatic tissue found in the small intestine.

H3. Answer these questions about the activities that occur in the small intestine.

a. *(Protein? Fat?)* digestion first occurs in the small intestine.

b. Define absorption.

c. Intestinal juice has a pH of _____ and contains enzymes produced by small intestinal
cells.

d. Two types of movement occur in the small intestine. The major movement is called

_____ and the other is referred to as peristalsis.

e. Briefly discuss how these two movements differ.

f. The end products of carbohydrate digestion are _____ ,

_____ , and _____ .

I. Large Intestine (pages 732–738)

I1. Answer the following questions about the large intestine.

a. The large intestine is structurally divided into four principal regions:

_____, _____, _____, and

_____ _____.

b. In order for chyme to move from the terminal portion of the small intestine into the large

intestine, the _____ _____ must relax.

c. At the origin of the large intestine is a blind pouch called the _____.

What structure attaches (inferiorly) to it? _____ _____

d. The total length of the large intestine is about _____ meters.

I2. Match the structure of the large intestine with the correct description.

A	anus	LCF	left colic flexure
AC	ascending colon	R	rectum
CN	anal canal	RCF	right colic flexure
CO	anal columns	SC	sigmoid colon
DC	descending colon	TC	transverse colon

a. _____ Normally closed except during elimination of waste.

b. _____ Retroperitoneal structure, on the right side of the abdomen.

c. _____ Crosses the abdomen from the right to the left side.

d. _____ Longitudinal folds that contain a network of arteries and veins.

e. _____ Passes downward to the level of the iliac crest.

f. _____ Curve located beneath the liver.

g. _____ Begins near the left iliac crest, projects toward the midline, and ends at the rectum.

h. _____ Curve located beneath the spleen.

i. _____ Last 20 cm of the GI tract; located anterior to the sacrum and coccyx.

j. _____ Terminal 2 to 3 cm of the rectum.

I3. Answer these questions about the large intestine histology.

a. The presence of numerous _____ cells helps to lubricate the colonic contents as they pass through the colon.

b. How does the longitudinal musculature of the large intestine differ from that of other portions of the GI tract? What structures are formed by this variation?

I4. Complete the following questions about the activities of the large intestine.

a. Contrast the three types of movement characteristic of the large intestine.
   1. Haustral churning

   2. Peristalsis

   3. Mass peristalsis

b. Explain how chemical digestion in the large intestine differs from that in other regions of the GI tract.

c. The chyme remains in the large intestine _____ to _____ hours and becomes a solid or

semisolid mass referred to as _____.

d. The absorption of water from the large intestine is greatest in the *(cecum and ascending colon? transverse and descending colon?)*.

e. Identify the regions of the large intestine in Figure LG 24.1. Draw arrows to indicate the direction of movement of the intestinal contents.

I5. Describe the process of defecation.

I6. Complete the table below, relating to the type of chemical and mechanical digestion that occurs in the stomach and the small and large intestines.

	Chemical	Mechanical
a. Stomach		
b. Small intestine		
c. Large intestine		

J. Developmental Anatomy of the Digestive System   (pages 739–740)

J1. Approximately the _____ day after fertilization the primitive gut forms.

J2. An anterior depression consisting of ectoderm, the _____, will develop into the oral cavity, while a posterior depression of ectoderm, the

_____, will form the anus.

K. Aging and the Digestive System   (pages 740–741)

K1. As a person ages, you would expect to find a(n) *(decrease? increase?)* in secretory mechanisms and a(n) *(decrease? increase?)* in motility.

K2. Name several pathologies of accessory structures that are associated with aging.

L. Applications to Health (Refer to the *Application to Health* companion to answer Part L questions.)

L1. Briefly explain the mechanism of tooth decay (dental caries).

L2. Match the disorder with its description.

AN	anorexia nervosa	H	hepatitis
B	bulimia	O	obesity
Ci	cirrhosis	PD	periodontal disease
D	diarrhea	PU	peptic ulcer
GS	gallstones		

a. _____ Characterized by overeating at least twice a week followed by purging by self-induced vomiting, strict dieting or fast, vigorous exercise, or use of laxatives or diuretics.

b. _____ Characterized by inflammation and degeneration of the gingivae, alveolar bone, periodontal ligaments, and cementum.

c. _____ Chronic disorder characterized by self-induced weight loss, body image and perceptual disturbances, and physiological changes that result from nutritional depletion.

d. _____ An inflammation of the liver caused by viruses, drugs, or chemicals.

e. _____ A body weight more than 20% above a desirable standard due to an excessive accumulation of fat.

f. _____ A craterlike lesion in a GI tract membrane that develops in areas of the GI tract exposed to acidic gastric juices.

g. _____ Stem from the fusion of crystals of cholesterol in bile, leading to the obstruction of the cystic or common bile duct.

h. _____ An increase in frequency, volume, and fluid content of the stool.

i. _____ A distorted or scarred liver as a result of chronic inflammation.

L3. Contrast the causes of the following:

a. Hepatitis A

b. Hepatitis B

ANSWERS TO SELECT QUESTIONS

A1. (a) Mouth; (b) pharynx; (c) esophagus; (d) stomach; (e) small intestine; (f) large intestine; (g) teeth; (h) tongue; (i) salivary glands; (j) liver; (k) gallbladder; (l) pancreas.

B1. (a) DF; (b) P; (c) D; (d) I; (e) A; (f) S.

B3. A—Parotid gland, B—pharynx, C—esophagus, D—liver, E—gallbladder, F—duodenum, G—colon, ascending, H—cecum, I—vermiform appendix, J—mouth, K—sublingual and submandibular glands, L—stomach, M—pancreas, N—colon, transverse, O—jejunum, P—colon, descending, Q—ileum, R—colon, sigmoid, S—rectum, T—anus.

C1. (a) Mucosa, 3; (b) lamina propria; (c) simple, secretion, absorption; (d) submucosal plexus; (e) circular, longitudinal; (f) visceral peritoneum; (g) ascites; (h) organs lie on the posterior abdominal wall covered by the peritoneum.

C2. (a) A—Longitudinal muscle, B—circular muscle, C—muscularis mucosae, D—lamina propria, E—epithelium, F—villus, G—gland in submucosa; (b) 1. mucosa, 2. submucosa, 3. muscularis, 4. serosa.

C3. (a) MC; (b) FL; (c) GO; (d) MS; (e) LO.

D1. (a) U; (b) LB; (c) VS; (d) F; (e) LF; (f) V.

D2. Palatoglossal, palatopharyngeal.

D3. (a) I; (b) E.

D4. (a) Filiform—contain no taste buds; (b) fungiform; (c) circumvallate.

D5. (a) Buccal, parotid, submandibular, sublingual; (b) moistens, lubricates, dissolves, cleanses, begins digestion; (c) sympathetic, vasoconstriction; (d) salivary amylase.

D6. Anchors the teeth in position; acts as a shock absorber when chewing.

D7. (a) CM; (b) C; (c) E; (d) R; (e) AF; (f) N; (g) RC; (h) D; (i) P.

D8. V.

D9. 6th, 20, 32.

E1. Mastication, bolus.

E2. Naso.

E3. Deglutition.

E4. Posterior, esophageal hiatus.

E6. Peristalsis.

E7. Lower esophageal sphincter.

E8.

	Mechanical	Chemical
Mouth	**Mastication**	Initiate starch digestion
Pharynx	**Deglutition**	**None, purely transportation**
Esophagus	Peristalsis	**None, purely transportation**

F1. Cardia > fundus > body > pylorus.

F2. (a) Has gastric pits lined with exocrine cells; (b) 3 layers: inner—oblique, middle—circular, outer—longitudinal.

F3.

Cell Name	Function	Secretion
a. Chief (zymogenic)	**Produces digestive enzymes**	**Pepsinogen and gastric lipase**
b. **Mucous**	Lubricates and protects	**Mucus**
c. **Parietal**	**Activates pepsinogen; needed to absorb Vit. B_{12}**	Hydrochloric acid intrinsic factor
d. G cells	**Stimulates HCl and pepsinogen secretions, contracts lower esophageal sphincter, increases motility of stomach, and relaxes the pyloric sphincter**	**Stomach gastrin**

F4. Mixing waves, chyme.

F5. (a) Protein, limited; (b) 2, 6, carbohydrates, protein, fats.

G1. (a) Posterior, greater, head; (b) pancreatic duct; (c) pancreatic islets, G—alpha cells, I—beta cells, SS—delta cells, PP—F-cells; (d) alkaline, sodium; (e) pancreatic amylase, protein.

G2. (a) Hypochondriac, epigastric; (b) ligamentum teres, umbilicus; (c) hepatocytes, central, sinusoids, stellate reticuloendothelial.

G3. BC > RLHD > CHD > CD > CBD.

G5. Vitamins A, B_{12}, D, E, K; iron and copper.

G6. Hepatic artery, hepatic portal vein.

G7. (a) Submucosa; (b) 10; (c) celiac, vagus.

G8. A—right hepatic duct, B—left hepatic duct, C—common bile duct, D—hepatopancreatic ampulla, E—common hepatic duct, F—sphincter of the hepatopancreatic ampulla.

H1. (a) Ileum, duodenum, jejunum; (b) mucosa, submucosa; (c) intestinal juice, alkaline; (d) villi, microvilli; (e) capillary, lacteal.

H3. (a) Fat; (c) 7.6; (d) segmentation; (f) glucose, galactose, fructose.

I1. (a) Cecum, colon, rectum, anal canal; (b) ileocecal sphincter; (c) cecum, vermiform appendix; (d) 1.5.

I2. (a) A; (b) AC; (c) TC; (d) CO; (e) DC; (f) RCF; (g) SC; (h) LCF; (i) R; (j) CN.

I3. (a) Goblet (mucus).

I4. (b) bacterial rather than enzymatic; (c) 3, 10, feces; (d) cecum and ascending colon.

I6.

	Chemical	Mechanical
a. Stomach	**Protein digestion—enzymatic (pepsin)**	**Mixing waves—maceration**
b. Small intestine	**Protein, carbohydrate, and fat digestion—enzymatic**	**Segmentation and peristalsis**
c. Large intestine	**Bacterial—not enzymatic**	**Haustral churning and peristalsis**

J1. Fourteenth (14th).

J2. Stomodeum, proctodeum.

K1. Decrease, decrease.

L2. (a) B; (b) PD; (c) AN; (d) H; (e) O; (f) PU; (g) GS; (h) D; (i) Ci.

SELF QUIZ

Choose the one best answer to the following questions.

1. In which digestive structure is protein digestion initiated?

   A. mouth
   B. esophagus
   C. stomach
   D. small intestine
   E. large intestine

2. Which structure secretes a substance that aids in fat digestion?

   A. salivary glands
   B. large intestine
   C. esophagus
   D. liver
   E. A and C are both correct

3. The substance that attaches the root of the tooth to the periodontal ligament is (the)

   A. enamel
   B. dentin
   C. cementum
   D. pulp
   E. apical canal

4. What structures are located on the posterior surface of the tongue and form an inverted V shape?

   A. filiform papillae
   B. fungiform papillae
   C. circumvallate papillae
   D. dermal papillae
   E. none of the above answers are correct

5. Fingerlike projections that line the small intestine and function to increase the surface area are called

   A. mucosa
   B. villi
   C. lacteals
   D. gastric pits
   E. capillaries

6. Intestinal juice has a pH of approximately _____.

   A. 2.6
   B. 3.6
   C. 4.6
   D. 6.6
   E. 7.6

7. The region of the stomach closest to the duodenum is called the _____.

   A. fundus
   B. cardia
   C. body
   D. pylorus
   E. jejunum

8. The removal of the gallbladder would cause a difficulty in

   A. protein digestion
   B. carbohydrate digestion
   C. fat digestion
   D. carbohydrate absorption
   E. B and D are both correct answers

9. The salivary glands located inferior and anterior to the ears, between the skin and the masseter muscle, are

   A. submandibular
   B. parotid
   C. sublingual
   D. buccal
   E. A and C are both correct answers

10. The peritoneal structure that binds the small intestine to the posterior abdominal wall is the

    A. falciform ligament
    B. mesocolon
    C. mesentery
    D. lesser omentum
    E. ligamentum teres

11. Which of the following is NOT an accessory organ?

    A. teeth
    B. tongue
    C. esophagus
    D. liver
    E. gallbladder

Arrange the answers in the correct order.

12. From origin to end. _____ _____ _____ _____ _____

    A. common bile duct
    B. common hepatic duct
    C. right/left hepatic duct
    D. bile capillaries
    E. hepatopancreatic ampulla

13. From the deepest to the most superficial. _____ _____ _____ _____
    A. submucosa
    B. mucosa
    C. serosa
    D. muscularis

14. Movement of chyme. _____ _____ _____ _____ _____
    A. cecum
    B. pylorus
    C. jejunum
    D. duodenum
    E. ileum

Answer (T) True or (F) False to the following questions.

15. _____ Starch digestion is initiated in the stomach.

16. _____ No digestion occurs in the esophagus.

17. _____ Failure of the lower esophageal sphincter to relax normally as food approaches it is referred to as achalasia.

18. _____ Occult blood refers to hidden blood in the feces.

19. _____ Meat and meat by-products are an excellent source of fiber (bulk or roughage).

20. _____ The terminal portion of the small intestine is the cecum.

21. _____ The three longitudinal muscular bands of the large intestine are called the haustra.

Fill in the blanks.

22. The _____ _____ is the largest peritoneal fold in the body.

23. Varicosities of the rectal veins are known as _____.

24. _____ refers to an inflammation of the intestines.

25. The diversion of the fecal stream through a surgical "stoma" that is affixed to the exterior of the abdominal wall

    is called a _____.

26. The _____ are lymphatic vessels located in the villi of the small intestine.

27. The endoscopic examination of the stomach is called _____.

ANSWERS TO THE SELF QUIZ

1. C	10. C	19. F
2. D	11. C	20. F
3. C	12. D, C, B, A, E	21. F
4. C	13. B, A, D, C	22. Greater omentum
5. B	14. B, D, C, E, A	23. Hemorrhoids
6. E	15. F	24. Enteritis
7. D	16. T	25. Colostomy
8. C	17. T	26. Lacteals
9. B	18. T	27. Gastroscopy

The Urinary System

SYNOPSIS

In the last chapter you journeyed through the system responsible for the digestion and absorption of nutrients into the body. These food substances, after assimilation, are utilized within the cells in a variety of chemical reactions. It is the metabolic by-products of these reactions—carbon dioxide, excess water, heat, plus toxic ammonia and urea—that comprise the body's waste products. In addition, sodium, chloride, sulfate, phosphate, and hydrogen ions tend to accumulate in excess of the body's needs. Elimination of the toxic materials and excess materials is essential to the maintenance of homeostatic balance. It is the primary function of the **urinary system** to help maintain this balance by controlling the composition and volume of the blood.

The urinary system consists of two **kidneys**, two **ureters**, one **urinary bladder**, and a single **urethra**. Located within each kidney are approximately 1,000,000 **nephrons.** The nephron is the functional unit of the kidney, and it is responsible for the regulation of the composition and volume of the blood.

In addition to the aforementioned roles, the urinary system is of utmost importance in the process of erythropoiesis, control of blood pH, regulation of blood pressure, and activation of vitamin D.

TOPIC OUTLINE AND OBJECTIVES

A. Introduction and the Kidneys

1. Identify the external and internal gross anatomy of the kidneys.
2. Describe the blood and nerve supply to the kidneys.
3. Detail the histology of the nephron.
4. Define the juxtaglomerular apparatus.
5. Explain the physiology of the nephron and urine formation.

B. Ureters

6. Describe the structure, histology, physiology, and blood and nerve supply of the ureters.

C. Urinary Bladder

7. Describe the structure, histology, physiology, and blood and nerve supply of the urinary bladder.

D. Urethra

8. Describe the structure, histology, and physiology of the urethra.

E. Developmental Anatomy of the Urinary System

F. Aging and the Urinary System

G. Applications to Health

9. Discuss the etiology and symptomatology of glomerulonephritis (Bright's disease), nephrotic syndrome, polycystic disease, renal failure, urinary tract infections (URI), and urinary bladder cancer.

I. Key Medical Terms Associated with the Urinary System

SCIENTIFIC TERMINOLOGY

Find an anatomical sample word for each prefix and suffix:

Prefix/Suffix	Meaning	Sample Word
azo-	nitrogen containing	
corpus-	body	
cyst-	bladder	
-emia	condition of blood	
glomus-	ball	
litho-	stone	
nephros-	kidney	
oligo-	scanty	
podos-	foot	
pyelo-	pelvis of kidney	
-uria	urine	

A. Introduction and the Kidneys (pages 748–761)

A1.   List the three functions of the kidneys.

a.

b.

c.

A2.   Answer these questions about the external and internal anatomy of the kidneys.

a.   The term _____ indicates that the kidneys are located behind the parietal peritoneum of the abdominal cavity.

b.   What structure would cause the right kidney to be slightly lower than the left kidney?

_____

c.   Nerves and blood and lymphatic vessels enter and exit the kidney through the renal

_____. This structure is the entrance to a cavity called the renal *(pelvis? sinus?)*.

d.   The outer, reddish area of a frontally sectioned kidney is called the

_____ and the inner area is referred to as the _____.

e.   Within the renal medulla are 8 to 18 cone-shaped structures called renal *(papillae? pyramids?)*.

f.   The parenchyma of each kidney consists of approximately 1 million microscopic units

called _____.

g.   The *(major? minor?)* calyces drain urine into the renal pelvis and out through the ureter into the urinary bladder.

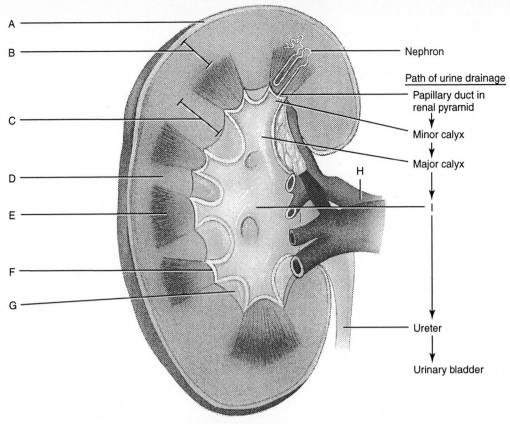

Diagram of frontal section of right kidney

Figure LG 25.1 Frontal section of right kidney.

A3. Match the connective layer of the kidney with its description.

| AC adipose capsule | RC renal capsule | RF renal fascia |

a. _____ This is the outermost layer.

b. _____ Serves as a barrier against trauma and maintains the shape of the kidney.

c. _____ Protects the kidney from trauma and holds the kidney firmly in place.

d. _____ This is the innermost layer.

e. _____ Anchors the kidney to its surrounding structures and the abdominal wall.

f. _____ This is the middle layer.

A4. Refer to Figure LG 25.1.

a. Identify structures A–J, which are parts of the kidney and urinary system.

_____ renal cortex   _____ renal pyramid

_____ renal medulla   _____ renal column

_____ renal papilla   _____ renal capsule

_____ renal pelvis   _____ renal artery

_____ renal sinus   _____ renal vein

A5. Approximately _____ ml of blood pass through the kidney each minute. This is about *(20–25%? 25–30%?)* of the resting cardiac output.

A6. After entering the kidney, the renal artery begins to divide. List, in order, the five branches that the blood must travel in prior to reaching the glomerulus.

a.

b.

c.

d.

e.

A7. Answer these questions about the blood and nerve supply of the kidney.

a. The *(peritubular capillaries? vasa recta?)* are associated only with a juxtamedullary nephron and not with a cortical nephron.

b. The renal plexus of the *(parasympathetic? sympathetic?)* division of the autonomic nervous system supplies the kidneys.

A8. The functional unit of the kidney is the _____. It is composed of two

portions: a _____ _____ and a

_____ _____. One component of the latter structure is a capillary network called the *(glomerulus? renal pyramid?)*.

A9. Using the terms below, trace the flow of the filtrate from the renal corpuscle (RC) to the papillary ducts (PD).

AL	ascending limb of nephron	DL	descending limb of nephron
CD	collecting ducts	LN	loop of the nephron
DC	distal convoluted tubule	P	proximal convoluted tubule

Renal corpuscle → (a) _____ → (b) _____ → (c) _____ →

(d) _____ → (e) _____ → (f) _____ → papillary ducts.

A10. Differentiate between the following terms.

a. Cortical nephron

b. Juxtamedullary nephron

A11. Substances filtered out of the blood pass through three structures that form the endothelial–capsular membrane. Write, in correct order, the three layers of this membrane that the filtered substances must pass through.

a. First:

b. Second:

c. Third:

A12. The specialized epithelial cells tht cover the glomerular capillaries are called

_____. Each cell has footlike processes referred to as

_____, which help to form spaces called the *(slit membrane? filtration slits?)*.

A13. Refer to Figure LG 25.2 and identify the following structures.

_____ afferent arteriole   _____ interlobular artery

_____ arcuate artery   _____ interlobular vein

_____ arcuate vein   _____ papillary duct

_____ collecting duct   _____ peritubular capillaries

_____ descending limb of the loop of Henle   _____ proximal convoluted tubule

_____ distal convoluted tubule   _____ renal papilla

_____ efferent arteriole   _____ thick ascending limb of the loop of Henle

_____ glomerulus   _____ thin ascending limb of the loop of Henle

_____ glomerular capsule   _____ vasa recta

A14. Answer these questions about the juxtaglomerular apparatus (JGA).

a. The JGA is located *(near? within?)* the glomerulus.

b. Some of the smooth muscle cells of the afferent arteriole have *(long? round?)* nuclei and their cytoplasm contains *(myofibrils? granules?)*. Such cells are called

_____ _____.

c. The cells of the distal ascending limb of the loop of Henle are *(shorter and more widely spread out? taller and more crowded together?)* than normal cells of the tubules. These

cells are known as the _____ _____.

d. What is the function of the JGA?

A15. List the three principal processes involved in urine formation.

_____ _____ _____

A16.

a. Approximately _____% of the filtrate moves from the tubules into the peritubular capillaries or vasa recta in a process called tubular reabsorption.

b. List four substances passed from the blood into the filtrate during tubular secretion.

_____ _____ _____

_____

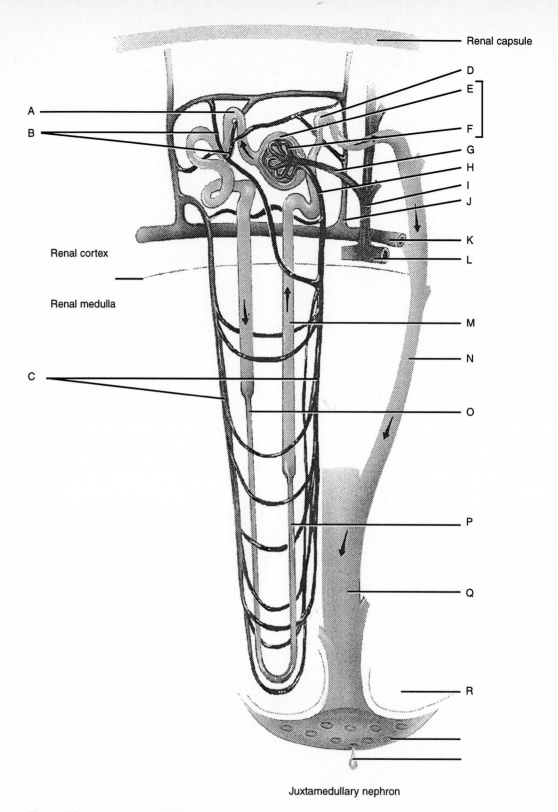

Renal capsule

A

B

Renal cortex

Renal medulla

C

D
E
F
G
H
I
J
K
L

M

N

O

P

Q

R

Juxtamedullary nephron

Figure LG 25.2   Juxtamedullary nephron.

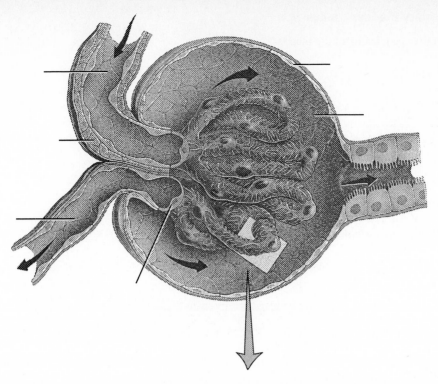

Figure LG 25.3 Parts of renal corpuscle (internal view) with endothelial–capsular membrane highlighted.

A17. Refer to Figure LG 25.3 and color and label the following structures. Be sure to match your color with the color code ovals.

O afferent arteriole
O capsular space
O efferent arteriole
O endothelium of the glomerulus
O juxtaglomerular cells
O parietal layer of the glomerular capsule

B. Ureters   (page 761)

B1. The ureters are approximately _____–_____ cm in length and enter the urinary bladder *(medially? laterally?)* from the posterior aspect.

B2. Since the ureters have no anatomical valves, how is the backflow of urine from the bladder prevented?

B3. There are _____ coats that form the wall of the ureters. Name the coat responsible for

peritalsis. _____

C. Urinary Bladder (page 762)

C1. Give the location of the bladder

a. in males:

b. in females:

C2. The *(internal? external?)* urethral orifice lies in the anterior corner of the trigone and is

surrounded by a circular layer of smooth muscle called the _____

urethral _____.

C3. What type of epithelium lines the bladder? _____

C4. The third coat of the bladder (the most superficial) is a muscular one called the

_____ _____, consisting of *(2? 3? 4?)* layers.

C5. Answer these questions about bladder physiology.

a. The act of urinating, which is also known as _____, occurs when the
amount of urine in the bladder exceeds *(200–300 ml? 200–400 ml?).*

b. _____ receptors in the wall of the bladder transmit impulses to the spinal
cord. These impulses, by way of sensory tracts to the cortex, initiate a conscious desire to
expel urine.

D. Urethra (pages 764–765)

D1. Describe the location of the female urethra.

D2. List the three structures that the male urethra passes through.

_____ _____ _____

D3. The wall of the female and male urethra consists of _____ coats.

D4. How does the mucosa of the prostatic urethra differ from that of the membranous and the
spongy urethra?

E. Developmental Anatomy of the Urinary System (page 765)

E1. The first kidney is called the _____, and the duct associated with it

drains into the _____.

E2. The _____ that replaces the first kidney begins to degenerate at

approximately the _____ week.

E3. What developmental structure gives rise to the urinary bladder? _____

_____

F. Aging and the Urinary System   (page 765)

F1. Name two common urinary disorders associated with aging.

a. _____

b. _____

F2. By age 70 the filtering mechanism of the kidneys is only about _____ of what it was at
age 40.

G. Applications to Health (Refer to the *Applications to Health* companion to answer Part G questions.)

G1. Match the urinary disorder with the correct description.

C	cystitis	PD	polycystic kidney disease
G	glomerulonephritis	P	pyelitis
N	nephrotic syndrome		

a. _____ Protein in the urine due to increased permeability of the endothelial–capsular
membrane.

b. _____ Inflammation of the urinary bladder involving the mucosa and submucosa.

c. _____ One of the most common inherited disorders of the kidneys; characterized by
deformed nephrons with cystlike dilations along their course.

d. _____ An inflammation of the kidney involving the glomeruli. It is commonly caused by
an allergic reaction to toxins given off by streptococci bacteria.

e. _____ An inflammation of the renal pelvis and its calyces.

G2. _____ renal failure is a progressive, generally irreversible decline in
the glomerular filtration rate.

G3. Match these terms with their definitions.

A	anuria	O	oliguria

a. _____ Daily urinary output less than 250 ml.

b. _____ Daily urinary output less than 50 ml.

G4. Every year nearly (*12,000? 120,000?*) people die from urinary bladder cancer. Its
incidence has increased about (*36% 46%*) over the past ten years.

ANSWERS TO SELECT QUESTIONS

A1. (a) Regulation of composition and volume of the blood; (b) help regulate blood pressure; (c) contribute to metabolism.

A2. (a) Retroperitoneal; (b) liver; (c) hilus, sinus; (d) cortex, medulla; (e) pyramids; (f) nephrons; (g) major.

A3. (a) RF; (b) RC; (c) AC; (d) RC; (e) RF; (f) AC.

A4. A—Renal capsule, B—renal cortex, C—renal medulla, D—renal column, E—renal pyramid, F—renal papilla, G—renal sinus, H—renal artery, I—renal pelvis, J—renal vein.

A5. 1200, 20–25%.

A6. (a) Segmental arteries; (b) interlobar arteries; (c) arcuate arteries; (d) interlobular arteries; (e) afferent arterioles.

A7 (a) Vasa recta; (b) sympathetic.

A8. Nephron, renal tubule, renal corpuscle, glomerulus.

A9. (a) P; (b) DL; (c) LN; (d) AL; (e) DC; (f) CD.

A11. (a) Endothelium of glomerulus; (b) basement membrane of glomerulus; (c) filtration slits in podocytes.

A12. Podocytes, pedicels, filtration slits.

A13. A—Proximal convoluted tubule, B—peritubular capillaries, C—vasa recta, D—distal convoluted tubule, E—glomerular capsule, F—glomerulus, G—afferent arteriole, H—efferent arteriole, I—interlobular artery, J—interlobular vein, K—arcuate vein, L—arcuate artery, M—thick ascending limb of the loop of Henle, N—collecting duct, O—descending limb of the loop of Henle, P—thin ascending limb of the loop of Henle, Q—papillary duct, R—renal papilla.

A14. (a) Near; (b) round, granules, juxtaglomerular cells; (c) taller and more crowded together, macula densa; (d) helps regulate blood pressure by secreting an enzyme that starts a sequence of reactions that raise the blood pressure and the rate of blood filtration by the kidneys.

A15. Glomerular filtration, tubular reabsorption, tubular secretion.

A16. (a) 99; (b) hydrogen ions, ammonia, creatine, certain drugs.

B1. 25–30, medially.

B2. Pressure in the urinary bladder compresses the ureteral openings.

B3. 3, (second) muscularis.

C1. (a) Anterior to the rectum; (b) anterior to the vagina and inferior to the uterus.

C2. Internal, internal, sphincter.

C3. Transitional.

C4. Detrusor muscle, 3.

C5. (a) Micturition, 200–400, (b) stretch.

D2. Prostate gland, urogenital diaphragm, penis.

D3. Two.

E1. Pronephros, cloaca.

E2. Mesonephros, sixth.

E3. Urogenital sinus.

F1. (a) urinary tract infection; (b) kidney inflammation.

F2. One-half.

G1. (a) N; (b) C; (c) PD; (d) G; (e) P.

G2. Chronic.

G3. (a) O; (b) A.

G4. 12,000, 36%.

SELF QUIZ

Choose the one best answer to the following questions.

1. The functional unit of the kidney is the

   A. renal pyramid
   B. nephron
   C. minor calyx
   D. renal column
   E. renal pelvis

2. Filtrate leaving the ascending limb of the loop of the nephron would next enter the

   A. glomerulus
   B. proximal convoluted tubule
   C. distal convoluted tubule
   D. collecting duct
   E. papillary duct

3. The renal arteries transport approximately _____ ml of blood to the kidneys every minute.

   A. 1000
   B. 1100
   C. 1200
   D. 1300
   E. 1400

4. Tubular reabsorption occurs primarily via the

   A. glomerulus
   B. efferent arteriole
   C. peritubular capillaries and vasa recta
   D. renal vein
   E. collecting ducts

Arrange the answers in the correct order.

5. From outer to inner layer around the kidney.   ＿＿＿ ＿＿＿ ＿＿＿

   A. renal capsule
   B. renal fascia
   C. adipose capsule

6. From glomerular capsule to collecting duct.   ＿＿＿ ＿＿＿ ＿＿＿ ＿＿＿ ＿＿＿

   A. distal convoluted tubule
   B. ascending limb of the loop of Henle
   C. loop of the nephron
   D. descending limb of the loop of Henle
   E. proximal convoluted tubule

7. Pathway of the blood   ＿＿＿ ＿＿＿ ＿＿＿ ＿＿＿

   A. efferent arteriole
   B. afferent arteriole
   C. glomerulus
   D. peritubular capillaries and vasa recta

8. Pathway of the blood (arteries).   ＿＿＿ ＿＿＿ ＿＿＿ ＿＿＿ ＿＿＿ ＿＿＿ ＿＿＿

   A. efferent arterioles
   B. interlobar
   C. interlobular
   D. afferent arteriole
   E. arcuate
   F. segmental
   G. glomerulus

9. From the distal convoluted tubules to the major calyces.   ＿＿＿ ＿＿＿ ＿＿＿

   A. minor calyces
   B. collecting ducts
   C. papillary ducts

10. Order through which substances filtered by the kidneys must pass.   ＿＿＿ ＿＿＿ ＿＿＿ ＿＿＿

   A. filtration slits
   B. basement membrane of glomerulus
   C. slit membrane
   D. fenestrated capillary (endothelium)

11. Urine flow from superior to inferior.   ＿＿＿ ＿＿＿ ＿＿＿ ＿＿＿

   A. urethra
   B. kidney
   C. urinary bladder
   D. ureter

Answer (T) True or (F) False to the following questions.

12. _____ Salts present in the urine may solidify into stones called renal calculi.

13. _____ Approximately 80% of the filtrate is moved back into the circulation via tubular reabsorption.

14. _____ The portions of the cortex that extend between the renal columns are called the renal papillae.

15. _____ Nephroptosis refers to a kidney that is displaced inferiorly.

16. _____ The glomerulus together with the glomerular capsule constitute the renal corpuscle.

17. _____ A cystocele is a hernia of the urinary bladder.

18. _____ The loss of voluntary control over micturition is referred to as retention.

19. _____ Enuresis refers to excessive urine formation.

Fill in the blanks.

20. _____ epithelium lines the inner surface of the urinary bladder.

21. The notch on the medial surface of a kidney is called the renal _____.

22. The muscular coat of the bladder is called the _____ muscle.

23. The _____ _____ helps regulate blood pressure by secreting renin.

24. _____ is an inflammation of the bladder.

25. _____ refers to increased excretion of urine and _____ refers to painful urination.

ANSWERS TO THE SELF QUIZ

1. B
2. C
3. C
4. C
5. B, C, A
6. E, D, C, B, A
7. B, C, A, D
8. F, B, E, C, D, G, A
9. B, C, A

10. D, B, C, A
11. B, D, C, A
12. T
13. F
14. F
15. T
16. T
17. T
18. F

19. F
20. Transitional
21. Hilus
22. Detrusor
23. Juxtaglomerular apparatus
24. Cystitis
25. Diuresis, dysuria

The Reproductive Systems

CHAPTER
26

SYNOPSIS

One of the strongest urges within nature is the propagation of a species. Sexual reproduction is a mechanism that allows for the transmission of genetic information from one generation to the next. The reproductive system provides the organs and accessory structures necessary for this continuity of life.

The organs of the male and female reproductive systems are classified into functional groups. The **gonads** (testes and ovaries) are responsible for the production of the gametes: **sperm cells** and **ova.** In addition, the gonads secrete sex hormones which are important to the development and maintenance of the reproductive system. This is especially true when discussing the **ovarian** and **uterine (menstrual) cycles.** The **ducts** of the reproductive system conduct, receive, and store gametes, while the **accessory structures** produce materials that support the gametes.

TOPIC OUTLINE AND OBJECTIVES

A. Male Reproductive System

1. Define reproduction and classify the organs of the reproductive system by function.

2. Describe the structure and function of the scrotum.

3. Discuss the structure, histology, and functions of the testes.

4. Define spermatogenesis and spermiogenesis and explain the processes.

5. Define spermatozoa and describe their structure.

6. Describe the location, structure, histology, and functions of the ductus epididymis, ductus (vas) deferens, ejaculatory duct, and urethra.

7. Describe the location and function of the seminal vesicles, prostate gland, and bulbourethral glands.

8. Describe the external and internal anatomy of the penis and the physiology of an erection.

B. Female Reproductive System

9. Describe the location, histology, and functions of the ovaries.

10. Define oogenesis and describe the process.

11. Describe the location, structure, and functions of the uterine (Fallopian) tubes.

12. Describe the location, structure, and function of the uterus.

13. Describe the location, structure, and function of the vagina.

14. Describe the components of the vulva and explain their function.

15. Explain the structure and histology of the mammary glands.

C. Uterine and Ovarian Cycles

16. Explain the events and importance of the menstrual cycle.

D. Developmental Anatomy of the Reproductive Systems

E. Aging and the Reproductive Systems

F. Applications to Health

17. Explain the etiology and symptomatology associated with selected sexually transmitted diseases (STDs) such as gonorrhea, syphilis, genital herpes, chlamydia, trichomoniasis, and genital warts.

18. Describe the etiology and symptomatology associated with male disorders (testicular cancer, prostate dysfunction, impotence, and infertility) and female disorders (amenorrhea, dysmenorrhea, premenstrual syndrome (PMS), endometriosis, infertility, breast tumors, cervical cancer, and pelvic inflammatory disease (PID).

G. Key Medical Terms Associated with the Reproductive Systems

SCIENTIFIC TERMINOLOGY

Find an anatomical sample word for each prefix and suffix:

Prefix/Suffix	Meaning	Sample Word
a-	without	
acro-	top	
albus-	white	
caverna-	hollow	
circumcido-	to cut around	
culp-	vagina	
ejectus-	to throw out	
epi-	above	
gyneco-	woman	
homo-	same	
hyster-	uterus	
kryptos-	hidden	
lact-	milk	
meio-	less	
men-	month	
myo-	muscle	
-oon	egg	
-orchis	testis	
salpingo-	tube	
-zoon	life	

A. Male Reproductive System (pages 771–784)

A1. Refer to Figure LG 26.1 and complete the following questions.

a. Identify the structures labeled 1–8 in the diagram.

1. _____   5. _____

2. _____   6. _____

3. _____   7. _____

4. _____   8. _____

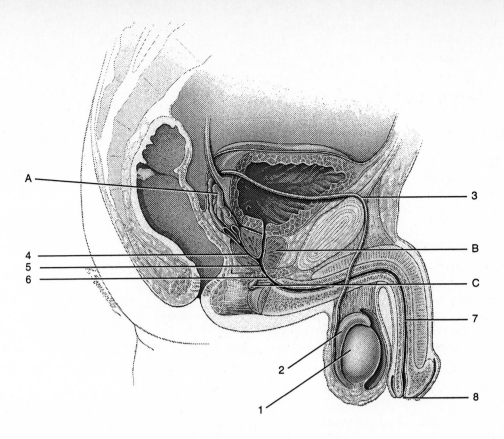

Figure LG 26.1 Male organs of reproduction seen in sagittal section.

b. The structures labeled A, B, and C represent the accessory sex glands. Identify each gland.

    A.

    B.

    C.

c. Using colored pencils, shade in the accessory sex glands, ducts, and gonads. (Note: be sure that your colors match the color code ovals.)

A2. Complete the following questions about the scrotum and testes.

a. The median ridge that appears to separate the scrotum into lateral portions is called the

    _____.

b. Wrinkling of the skin of the scrotum is created by the presence of the

    _____ muscle.

c. What effect does having the testes located in the scrotal sac have on their temperature?

d. What is the function of the cremaster muscle?

e. The testes originate in the posterior abdominal wall and descend through the

   _____ _____ (passageways in the anterior abdominal
   wall) into the scrotum.

f. An outpocketing of the peritoneum that partially covers the testes is called the

   _____ _____. Internal to this structure is a dense white

   fibrous capsule known as the _____ _____.

g. Each testis is divided into approximately *(100–200? 200–300?)* lobules, each containing

   one to three tightly coiled tubules called the _____ tubules.

h. Embedded between the spermatogenic cells are the _____ cells, which
   are responsible for the protection and nourishment of the sperm cells and for the secretion
   of the hormone *(FSH? inhibin?).*

i. The cells responsible for the secretion of the male sex hormone, testosterone, are the

   _____ _____.

j. Undescended testes is a condition referred to as _____.

A3. Moving from the basement membrane (BM) of the seminiferous tubule (ST) toward the
   lumen, place the following spermatogenic cells in their proper order.

ES	early spermatid	SG	spermatogonia
LS	late spermatid	SS	secondary spermatocyte
PS	primary spermatocyte	SZ	spermatozoon

   BM → (a) _____ → (b) _____ → (c) _____ → (d) _____ → (e) _____

   → (f) _____ → ST

A4. Answer these questions about spermatogenesis and spermiogenesis.

a. Mature sperm cells contain *(46? 23?)* chromosomes; therefore, they have a *(haploid?
   diploid?)* number of chromosomes.

b. In humans, spermatogenesis takes about _____ days.

c. Meiosis consists of two divisions: the first division is known as _____

   division and the second is referred to as _____ division.

d. DNA replication occurs *(prior to? after?)* the reduction division and results in the formation
   of 46 chromosomes with *(half? double?)* the DNA.

e. The exchange of genetic materials (genes) among maternal and paternal chromosomes is

   called _____ _____.

f. Ultimately, each primary spermatocyte will eventually produce *(1? 2? 3? 4?)* spermatids.

g. Define spermiogenesis

h. During spermiogenesis the spermatids are embedded among the sustentacular cells. The

release of a mature spermatozoon from a sustentacular cell is known as _____.

A5. Complete these questions about spermatozoa.

a. Spermatozoa are produced at a rate of about _____/day. Upon ejaculation they have a life
expectancy of about *(48? 72?)* hours.

b. The genetic material is located within the _____ of the spermatozoon.

c. The function of the acrosome is

d. Mitochondria are located in the _____ of the spermatozoon.

e. What is the function of the tail? _____

A6. Match the duct through which the spermatozoa travel, from the seminiferous tubules to
the ductus deferens, with its correct description.

B	body of the epididymis	R	rete testis
E	efferent ducts	S	straight tubules
H	head of the epididymis	T	tail of the epididymis

a. _____ Immediately follow the seminiferous tubules.

b. _____ Large, superior portion of the epididymis.

c. _____ "Network" of ducts in the testis.

d. _____ Transports sperm out of the testis into the ductus epididymis.

e. _____ The narrow midportion of the epididymis.

f. _____ At its distal end is the ductus deferens.

A7. If uncoiled, the ductus epididymis is approximately _____m in length. The free
surfaces of the cells that line the ductus epididymis contain long, branching microvilli

called _____.

A8. a. What is the function of the ductus epididymis?

b. How many days are required for sperm to complete their maturation?

A9. Answer these questions about the ductus deferens.

a. Another name for this duct is the *(ejaculatory? vas?)* deferens.

b. In order to enter the pelvic cavity, the ductus deferens penetrates the

_____ _____.

c. A vasectomy involves the removal of a portion of the ductus deferens within the

_____.

d. After a vasectomy is performed, the production of sperm *(ceases? continues?)*.

e. Explain why a vasectomy does not affect sexual desire and performance.

f. List the structures that constitute the spermatic cord.

_____   _____

_____   _____

_____   _____

g. A weakness of the abdominal wall with the protrusion of part of an organ in the inguinal

region is called an _____ _____.

A10. The joining of the ductus deferens and the duct from the seminal vesicle forms the

_____ duct.

A11. List the three parts of the male urethra.

a.

b.

c.

A12. Match the accessory gland with its description.

B bulbourethral glands	P prostate gland	S seminal vesicles

a. _____ Secrete an alkaline, viscous fluid that is rich in fructose, prostaglandins, and clotting proteins.

b. _____ Single gland about the size of a chestnut.

c. _____ Located on either side of the membranous urethra within the urogenital diaphragm.

d. _____ Lie posterior to and at the base of the urinary bladder anterior to the rectum.

e. _____ Secretes a milky fluid rich in acidic compounds and several enzymes.

f. _____ Neutralize the acid environment of the urethra and lubricate the end of the penis during intercourse.

A13. Complete this exercise about the semen.

a. The average number (range) of spermatozoa in an ejaculate is _____ million/ml. When

the number drops below _____ million/ml, the male is considered infertile.

b. Why are so many sperm cells required to fertilize a single ovum?

c. Semen has a slightly *(acidic? alkaline?)* pH, which neutralizes the acidic environment of the male urethra and female vagina.

A14. Answer these questions about the penis.

a. The two dorsolateral masses of the body of the penis are called the corpora

_____ penis. The small midventral mass is referred to as the corpus

_____ penis. All three masses consist of erectile tissue permeated by

_____ _____. Vascular changes within these structures

result in an _____.

b. The slightly enlarged distal end of the penis is called the _____

_____. It is covered by a loosely fitting _____ or
prepuce.

c. The removal of all or part of the prepuce is known as a _____.

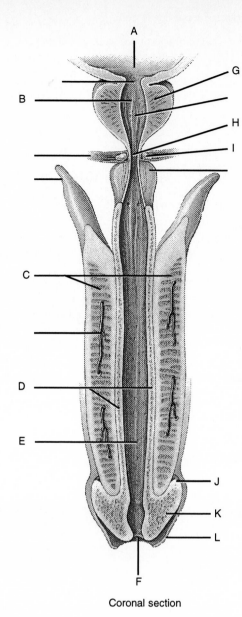

Coronal section

Figure LG 26.2   Internal structure of the penis.

d. Refer to Figure LG 26.2 and identify the following structures.

_____ bulbourethral (Cowper's) gland   _____ membranous urethra

_____ corona   _____ prepuce

_____ corpora cavernosa penis   _____ prostate gland

_____ corpus spongiosum penis   _____ prostatic urethra

_____ external urethral orifice   _____ spongy (cavernous) urethra

_____ glans penis   _____ urinary bladder

e. Refer to Figure LG 26.2 and label the non-lettered leader lines.

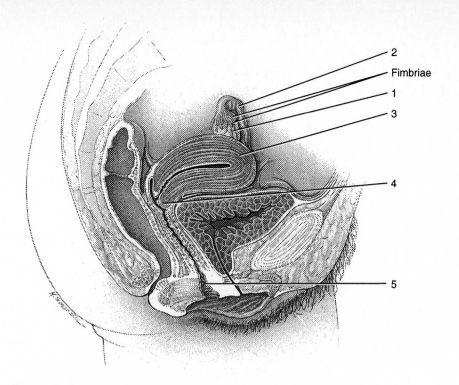

Figure LG 26.3 Female organs of reproduction.

B. Female Reproductive System   (pages 784–798)

B1. Answer the following questions pertaining to the female reproduction system.

a. Refer to Figure LG 26.3 and identify the structures labeled 1–5 in the diagram.

1. _____   4. _____

2. _____   5. _____

3. _____

b. Using colored pencils, shade the rectum (brown), urinary bladder (yellow), and the uterus (red).

c. The ovaries, which resemble *(unshelled almonds? walnuts?)*, are located in the pelvic cavity. They are held in place by ligaments. Name three of these structures.

1.

2.

3.

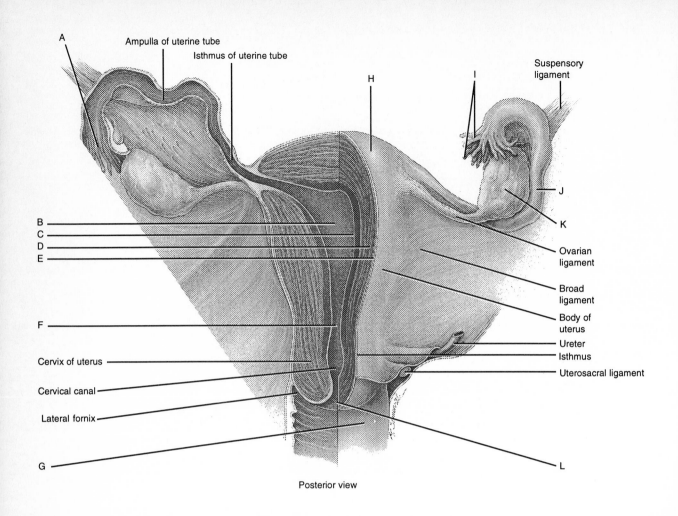

Figure LG 26.4 Uterus and associated structures.

d. Refer to Figure LG 26.4 and identify the following structures.

_____ endometrium   _____ myometrium

_____ external os   _____ ovary

_____ fimbriae of the uterine tube   _____ perimetrium

_____ fundus of uterus   _____ uterine cavity

_____ infundibulum of uterine tube   _____ uterine (Fallopian) tube

_____ internal os   _____ vagina

B2. Match the parts of the ovaries with the correct description.

C	corpus luteum	O	ovarian follicle
G	germinal epithelium	T	tunica albuginea
M	mature follicle		

a. _____ Oocytes in various stages of development and their surrounding, protective cells.

b. _____ Contains the remnants of an ovulated mature follicle.

c. _____ A whitish capsule of dense, irregular connective tissue.

d. _____ Covers the free surface of the ovary and is continuous with the mesothelium of the mesovarium.

e. _____ Relatively large, fluid-filled follicle that will soon rupture and expel a secondary oocyte.

B3. Complete these questions about oogenesis.

a. The final products of oogenesis are _____ haploid ovum and _____ haploid polar bodies.

b. At birth, there are approximately _____–_____ oogonia and primary oocytes in each ovary.

c. The _____ _____ is a clear glycoprotein layer that forms between the oocyte and granulosa cells as a follicle grows.

B4. Describe what a polar body is.

B5. Complete the following questions about the uterine tubes.

a. The uterine (Fallopian) tubes are also called the _____. At the distal end

of an uterine tube is a funnel-shaped opening called the _____, which is

surrounded by fingerlike projections called _____.

b. What is the function of these fingerlike projections?

c. The _____ is the widest portion of the uterine tube, while the short,

narrow, thick-walled portion that joins the uterus is called the _____.

d. What are the two functions of the uterine tubes?

1.

2.

B6. Complete the following questions about the uterus.

a. The uterus is situated between the _____ _____

(anteriorly) and the _____ (posteriorly).

b. The uterus is *(smaller? larger?)* when female sex hormones are low.

c. The three principal parts of the uterus are the _____,

_____, and _____

d. Contrast anteflexion and retroflexion of the uterus.

e. The _____ os is located at the junction of the isthmus with the cervical canal.

f. A procedure involving the removal and microscopic examination of a few cells from the

cervix is known as a _____ smear.

B7. Match the uterine ligament with the correct description.

B	broad ligaments	R	round ligaments
C	cardinal ligaments	U	uterosacral ligaments

a. _____ Extend from the uterus to a portion of the labia majora of the external genitalia.

b. _____ Attach the uterus to either side of the pelvic cavity.

c. _____ Connect the uterus to the sacrum.

d. _____ Extend below the bases of the broad ligaments between the pelvic wall and the cervix and vagina.

B8. Complete this exercise about the wall of the uterus.

a. The thickest layer of the uterine wall is the _____, which is made of *(skeletal? smooth?)* muscle fibers.

b. The _____ is the outer layer and is part of the visceral peritoneum.

c. The endometrium is a(n) *(avascular? vascular?)* layer that is divided into two layers. Which portion is shed during menstruation? Stratum *(functionalis? basalis?).*

B9. List the three functions of the vagina.

a.

b.

c.

B10. Define the following:

a. Rugae

b. Vaginal orifice

c. Hymen

B11. Match the following terms pertaining to the vulva with the correct description.

C	clitoris	P	prepuce
G	glans	V	vestibule
LMA	labia majora	VO	vaginal orifice
LMI	labia minora	VU	vulva
M	mons pubis		

a. _____ Collective designation for the female external genitalia.

b. _____ Cleft between the labia minora.

c. _____ Layer of skin formed at the point where the labia minora unite and cover the body of the clitoris.

d. _____ Cylindrical mass of erectile structure and nerves.

e. _____ Two folds of skin, devoid of pubic hair and fat, which have few sudoriferous glands.

f. _____ An elevation of adipose tissue covered by skin and coarse pubic hair.

g. _____ Are homologous to the scrotum.

h. _____ Exposed portion of the clitoris.

i. _____ Opening of the vagina; bordered by the hymen.

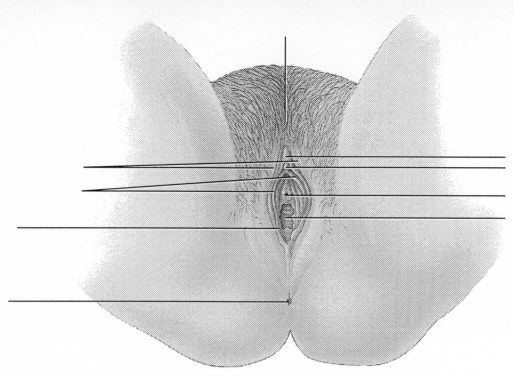

Figure LG 26.5   Perineum.

B12. Refer to Figure LG 26.5. Use the leader lines to label the following structures: anus, clitoris, hymen, labia majora, labia minora, mons pubis, prepuce, external urethral orifice, and vaginal orifice.

B13. Match the accessory glands with the descriptions.

L lesser vestibular	G greater vestibular	P paraurethral

a. _____ Located on either side of the vaginal orifice; homologous to the male bulbourethral glands.

b. _____ Have microscopic orifices that open into the vestibule.

c. _____ Located on either side of the urethral orifice; homologous to the male prostate gland.

B14. The borders of the perineum are the _____ _____

anteriorly, the _____ _____ laterally, and the

_____ posteriorly.

B15. Answer these questions about the mammary glands.

a. The mammary glands are modified _____ glands. Internally, each gland

consists of _____ to _____ lobes, separated by adipose tissue.

b. Clusters of milk-secreting cells of the mammary glands are known as

_____.

c. Breast size is determined by the amount of *(muscle? adipose?)* tissue in it.

d. The _____ refers to the pigmented area of skin surrounding the nipple.

e. _____ ligaments of the breast are responsible for support of the breasts.

f. Lactation describes breast milk secretion and ejection. The secretion is due largely to the

   hormone _____, and ejection occurs in the presence of

   _____.

B16. After milk is secreted it passes through several ducts to the nipple. Place the tubules/ducts
in their proper order from alveoli to nipple.

| LD | lactiferous ducts | LS | lactiferous sinuses |
| MD | mammary ducts | ST | secondary tubules |

Alveoli → _____ → _____ → _____ → _____ → nipple

B17. Refer to Figure LG 26.6 and identify the following structures.

_____ areola   _____ mammary duct

_____ deep fascia   _____ nipple

_____ fat in superficial fascia   _____ pectoralis major muscle

_____ intercostal muscles   _____ rib

_____ lactiferous duct   _____ secondary tubule

_____ lactiferous sinus   _____ suspensory ligament

_____ lobule containing alveoli

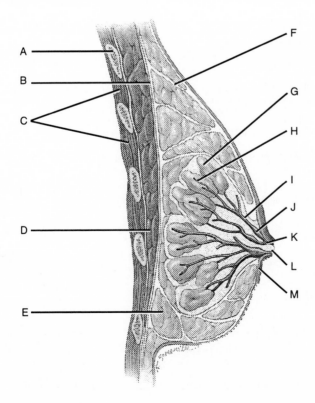

Sagittal section

Figure LG 25.6 Mammary glands.

C. Uterine and Ovarian Cycles (pages 798–800)

C1. Briefly contrast the menstrual and ovarian cycles.

C2. Match the hormone with the gland responsible for its secretion. (Note to the student: You may desire to review Chapter 22—The Endocrine System.)

E	estrogen	I	inhibin
F	follicle-stimulating hormone	L	luteinizing hormone
G	gonadotropin-releasing hormone	P	progesterone
		R	relaxin

anterior pituitary _____

hypothalamus _____

ovaries (corpus luteum) _____

placenta _____

C3. Using the abbreviations in question C2, match the hormone with the correct description. (Note: You may also refer to Tables 22.1 and 22.7.)

a. _____ Stimulates the release of FSH from the anterior pituitary gland.

b. _____ Initiates the development of the ovarian follicle.

c. _____ Brings about ovulation.

d. _____ Responsible for the development and maintenance of female reproductive structures and characteristics.

e. _____ Stimulates the release of LH from the anterior pituitary gland.

f. _____ Inhibits the release of FSH.

g. _____ Helps to dilate the uterine cervix to facilitate delivery.

h. _____ Converts the ovarian follicle into the corpus luteum.

i. _____ Prepares the endometrium for implantation and the mammary glands for milk secretion.

j. _____ High level of these hormones inhibits GnRH (two answers).

C4. Check your understanding of the menstrual cycle by answering the following questions.

a. The average duration of the menstrual cycle is _____ days. It is divided into four

   events: _____, _____, _____, and

   _____.

b. Menstruation is caused by the sudden *(increase? decrease?)* in estrogen and progesterone.

c. At birth each ovary contains about _____ primary follicles. (Note: See

   section on the ovary.) Toward the end of the menstrual phase about _____ of these follicles develop into secondary follicles.

d. As the secondary follicles continue to grow, one dominant follicle becomes the

_____ _____.

e. The preovulatory phase lasts from days _____ to _____ in a 28-day cycle. The

hormones _____ and _____ (from the anterior pituitary) stimulate the ovarian follicles to produce more estrogens, which in turn repair the *(perimetrium? endometrium?)*.

f. Just prior to ovulation, high levels of _____ and GnRH have a direct effect on the release of LH.

g. Following ovulation the mature follicle collapses. The blood clot that forms

within it is called the corpus _____. In time this clot is reabsorbed and

the corpus _____ forms, which secretes _____ ,

_____ , relaxin, and inhibin.

h. The last phase of the menstrual cycle is the _____ phase, which lasts

from days _____ to _____ in a 28-day cycle.

i. _____ is the dominant postovulatory hormone.

j. If fertilization and implantation do not occur, the corpus luteum degenerates into the

_____ _____. The absence of progesterone and estrogen *(promotes? inhibits?)* the output of GnRH, FSH, and LH secretions.

k. If fertilization and implantation occur, the corpus luteum is maintained by a hormone called

_____ _____ _____ (hCG), which

is produced by the _____.

D. Developmental Anatomy of the Reproductive Systems   (page 801)

D1. Primitive gonads for both sexes are present in the embryo. The *(presence? absence?)* of the Y chromosome determines the differentiation into a male.

D2. The prostate and bulbourethral glands are *(mesodermal? endodermal?)* outgrowths of the urethra.

D3. The external genitals of male and female embryos remain undifferentiated until about the *(7th? 8th? 9th)* week.

E. Aging and the Reproductive Systems   (pages 801–804)

E1. What changes are associated with the decreased synthesis of testosterone with aging?

E2. Uterine cancer peaks at about _____ years of age.

F. Applications to Health (Refer to the *Applications to Health* companion to answer Part F questions.)

F1. Match the sexually transmitted disease (STD) with the correct description.

C	chlamydia	GW	genital warts
G	gonorrhea	S	syphilis
GH	genital herpes	T	trichomoniasis

a. _____ An incurable disorder, commonly caused by the human papillomavirus (HPV).

b. _____ Infectious sexually transmitted disease that affects the mucous membranes of the urogenital tract, rectum, and occasionally the eyes. Caused by the *Neisseria* bacterium.

c. _____ Incurable disease, characterized by painful genital blisters on the prepuce, glans penis, and penile shafts in males and on the vulva and high up in the vagina in females.

d. _____ Caused by a flagellated protozoan; symptoms include a yellow vaginal discharge with an offensive odor and vaginal itching.

e. _____ Transmitted by a bacterium that cannot grow outside the body; principal symptom is urethritis in males and females, with possible spread through the female reproductive tract.

f. _____ Bacterial disease whose chief symptoms include chancres in the primary stage.

F2. Complete the following questions about male disorders.

a. Testicular cancer is one of the most common cancers seen in males ages _____ to

_____ .

b. A common disorder affecting one-third of all males over age 60, characterized by nocturia, hesitancy, and postvoid dribbling, is a condition called _____

_____ _____ .

c. Define impotence and male infertility. Are they the same condition?

F3. Answer the following questions about female disorders.

a. _____ is a condition characterized by the growth of endometrial tissue outside the uterus.

b. A common, benign breast tumor often seen in young women is a _____ .

c. Cervical cancer starts with _____ _____ , which is a change in size, shape, growth, and number of the cervical cells.

d. Pelvic inflammatory disease (PID) is most commonly caused by the bacterium that causes

_____ and _____ .

F4. List six factors that increase the risk of breast cancer.

a.

b.

c.

d.

e.

f.

ANSWERS TO SELECT QUESTIONS

A1. (a) 1—Testis, 2—epididymis, 3—ductus (vas) deferens, 4—ejaculatory duct, 5—prostatic urethra, 6—membranous urethra, 7—spongy urethra, 8—external urethral orifice; (b) A—seminal vesicle, B—prostate gland, C—bulbourethral (Cowper's) gland.

A2. (a) Raphe; (b) dartos; (c) lower than core body temperature; (d) elevates testes during sexual arousal and on exposure to cold; (e) inguinal canal; (f) tunica vaginalis, tunica albuginea; (g) 200–300, seminiferous; (h) sustentacular, inhibin; (i) interstitial endocrinocytes; (j) cryptorchidism.

A3. (a) SG; (b) PS; (c) SS; (d) ES; (e) LS; (f) SZ.

A4. (a) 23, haploid; (b) 74; (c) reduction, equatorial; (d) prior to, double; (e) crossing over; (f) 4; (h) spermiation.

A5. (a) 300,000,000, 48; (b) head; (c) contains enzymes to aid in penetration of the sperm cell into the secondary oocyte; (d) midpiece; (e) propels the sperm.

A6. (a) S; (b) H; (c) R; (d) E; (e) B; (f) T.

A7. 6, stereocilia.

A8. (a) Site of sperm maturation; (b) 10–14.

A9. (a) Vas; (b) inguinal canal; (c) scrotum; (d) continues; (f) ductus deferens, testicular artery, autonomic nerves, veins, lymphatic vessels, and the cremasteric muscle; (g) inguinal hernia.

A10. Ejaculatory.

A11. (a) Prostatic; (b) membranous; (c) spongy.

A12. (a) S; (b) P; (c) B; (d) S; (e) P; (f) B.

A13. (a) 50 to 150, 20; (b) a single sperm does not produce enough enzymes to dissolve the barrier surrounding the ovum. Fertilization requires the combined action of many sperm; (c) alkaline.

A14. (a) Cavernosa, spongiosum, blood sinuses, erection; (b) glans penis, foreskin; (c) circumcision; (d) A—urinary bladder, B—prostatic urethra, C—corpora cavernosa penis, D—corpus spongiosum penis, E—spongy (cavernous) urethra, F—external urethral orifice, G—prostate gland, H—membranous urethra, I—bulbourethral (Cowper's) gland, J—corona, K—glans penis, L—prepuce.

B1. (a) 1—ovary, 2—uterine (Fallopian) tube, 3—uterus fundus, 4—cervix, 5—vagina; (c) unshelled almonds, broad ligament, ovarian ligament, suspensory ligament; (d) A—infundibulum of uterine tube, B—uterine cavity, C—endometrium, D—myometrium, E—perimetrium, F—internal os, G—vagina, H—fundus of uterus, I—fimbriae of the uterine tube, J—uterine tube, K—ovary, L—external os.

B2. (a) O; (b) C; (c) T; (d) G; (e) M.

B3. (a) 1, 3; (b) 200,000, 2,000,000; (c) zona pellucida.

B5. (a) Oviducts, infundibulum, fimbriae; (b) help to carry the oocyte into the uterine tube; (c) ampulla, isthmus; (d) transports the oocyte toward the uterus, site of fertilization.

B6. (a) Urinary bladder, rectum; (b) smaller; (c) fundus, body, cervix; (d) internal; (f) Pap.

B7. (a) R; (b) B; (c) U; (d) C.

B8. (a) Myometrium, smooth; (b) perimetrium; (c) vascular, functionalis.

B9. (a) Passageway for menstrual flow; (b) passageway for childbirth; (c) receives semen from the penis during intercourse.

B11. (a) VU; (b) V; (c) P; (d) C; (e) LMI; (f) M; (g) LMA; (h) G; (i) VO.

B13. (a) G; (b) L; (c) P.

B14. Pubic symphysis, ischial tuberosities, coccyx.

B15. (a) Sudoriferous, 15, 20; (b) alveoli; (c) adipose; (d) areola; (e) suspensory (Cooper's); (f) prolactin, oxytocin.

B16. ST > MD > LS > LD.

B17. A—Rib, B—deep fascia, C—intercostal muscles, D—pectoralis major muscle, E—fat in superficial fascia, F—suspensory ligament, G—lobule containing alveoli, H—secondary tubule, I—mammary duct, J—lactiferous sinus, K—lactiferous duct, L—nipple, M—areola.

C2. Anterior pituitary: F, L; hypothalamus: G; ovaries: E, P, R, I; placenta: E, P.

C3. (a) G; (b) F; (c) L; (d) E; (e) E; (f) I; (g) R; (h) L; (i) P; (j) E, P.

C4. (a) 28–35, menstrual, preovulatory, ovulatory, postovulatory; (b) decrease; (c) 200,000–2,000,000, 20; (d) mature follicle; (e) 6–13, FSH, LH, endometrium; (f) estrogen; (g) hemorrhagicum, luteum, estrogen, progesterone; (h) postovulatory, 15–28; (i) progesterone; (j) corpus albicans, promotes; (k) human chorionic gonadotropin, placenta.

D1. Presence.

D2. Endodermal.

D3. 8th.

E2. 65.

F1. (a) GW; (b) G; (c) GH; (d) T; (e) C; (f) S.

F2. (a) 20, 35; (b) benign prostatic hyperplasia (BPH).

F3. (a) endometriosis; (b) fibroadenoma; (c) cervical dysplasia; (d) gonorrhea, chlamydia.

SELF QUIZ

Choose the one best answer to the following questions.

1. Follicle maturation and ovulation are controlled by_____.

   A. hGH and LH
   B. PRL and FSH
   C. estrogen and LH
   D. FSH and LH
   E. none of the above are correct

2. Fertilization most often occurs in the _____.

   A. ovary
   B. oviduct (uterine tube)
   C. uterus
   D. vagina
   E. external os

3. The portion of the sperm cell that contains numerous mitochondria, which provide energy for locomotion, is the

   A. flagellum
   B. acrosome
   C. head
   D. midpiece
   E. tail

4. The gland(s) which contribute to semen production is/are the (1) testis, (2) bulbourethral glands, (3) seminal vesicles, or (4) prostate gland.

   A. 1 only
   B. 2 only
   C. 3 only
   D. 1 and 3
   E. all of the above

5. Which hormone is associated with the maturation of an ovum?

   A. prolactin
   B. testosterone
   C. relaxin
   D. progesterone
   E. follicle-stimulating hormone

6. The principal hormone of the postovulatory phase is

   A. FSH
   B. LH
   C. relaxin
   D. progesterone
   E. estrogen

7. The female structure that is homologous to the penis is the _____.

   A. labia majora
   B. vagina
   C. clitoris
   D. hymen
   E. paraurethral glands

8. Which accessory gland in males secretes a fluid rich in acidic compounds and enzymes?

   A. testes
   B. seminal vesicles
   C. prostate gland
   D. bulbourethral glands
   E. Cowper's glands

9. A vasectomy involves the transection of which duct?

   A. seminiferous tubules
   B. efferent duct
   C. ductus deferens
   D. epididymis
   E. rete testis

10. Spermatogenesis results in the production of

    A. 4 spermatids, each $2n$
    B. 2 spermatids, each n
    C. 2 spermatids, each $2n$
    D. 4 spermatids, each n
    E. both B and D are correct answers

11. Spermatozoa are produced or mature at the rate of about _____ per day.

    A. 50,000,000
    B. 100,000,000
    C. 300,000,000
    D. 500,000,000
    E. 600,000,000

12. The enlarged distal region of the penis is called the _____.

    A. bulb
    B. cruca
    C. prepuce
    D. glans penis
    E. corona

13. Ovulation occurs (approximately) on the _____ day after the onset of menses.

   A. 9th
   B. 11th
   C. 14th
   D. 17th
   E. 22nd

14. Which portion of the uterus thickens from day 5 through day 23 of the normal uterine cycle?
   A. myometrium
   B. stratum basalis
   C. perimetrium
   D. stratum functionalis
   E. none of the answers are correct

15. Maturation of sperm cells occurs in the _____.

   A. prostate
   B. scrotum
   C. epididymis
   D. ejaculatory duct
   E. spongy urethra

Answer (T) True or (F) False to the following questions.

16. _____ The sustentacular cells produce and secrete testosterone.

17. _____ Amenorrhea refers to pain associated with menstruation.

18. _____ The corpus albicans secretes GnRH and LH.

19. _____ The structure of the mammary glands is formed primarily by adipose tissue.

Arrange the answers in the correct order.

20. Pathway of sperm from seminiferous tubules to the ejaculatory ducts.

   A. vas deferens
   B. straight tubule
   C. ductus epididymis
   D. rete testis
   E. efferent duct

   _____ _____ _____ _____ _____

21. From the deepest to the external (most superficial).   _____ _____ _____

   A. perimetrium
   B. endometrium
   C. myometrium

22. Pathway of milk from alveoli to the nipple   _____ _____ _____ _____

   A. lactiferous duct
   B. mammary duct
   C. secondary tubules
   D. lactiferous sinuses

Fill in the blanks.

23. _____ refers to a displaced urethral opening.

24. A procedure used to view the female pelvic cavity via the vagina is called a _____.

25. The removal of a uterine (Fallopian) tube is referred to as a _____.

ANSWERS TO THE SELF QUIZ

1. D	10. D	19. T
2. B	11. C	20. B, D, E, C, A
3. D	12. D	21. B, C, A
4. E	13. C	22. C, B, D, A
5. E	14. D	23. Hypospadias
6. D	15. C	24. Culdoscopy
7. C	16. F	25. Salpingectomy
8. C	17. F	
9. C	18. F	

Developmental Anatomy

SYNOPSIS

Your adventure is drawing to a close. In the last twenty-six chapters you have traveled throughout the body, visiting every system while discovering the intricacies of the tissues and the complexities of the cells. Despite the splendor of its design, you have also witnessed the myriad of disorders that affect this wondrous machine.

In the final leg of your journey you will have the opportunity to witness the development of a human being. **Developmental anatomy** is the study of the sequence of events from the fertilization of a secondary oocyte to the formation of an adult organism. In this chapter you will consider **fertilization, implantation, placental development, embryonic development, fetal growth, gestation, parturition**, and **labor**.

The birth of a baby has often been called a "miracle" and rightfully so. The miracle does not end at birth, however, for life itself is a miracle. Hopefully, your fantastic journey will leave you with a new appreciation for your own body and the many wonderful things it does.

TOPIC OUTLINE AND OBJECTIVES

A. Development During Pregnancy

1. Explain the processes associated with fertilization, morula formation, blastocyst development, and implantation.

B. Embryonic Development

2. Discuss the formation of the three primary germ layers, the embryonic membranes, placenta, and umbilical cord.
3. List the structures produced by the germ layers.

C. Fetal Growth

4. Describe the developmental changes associated with fetal growth.

D. Structural and Functional Changes During Pregnancy

5. Describe the anatomical and physiological changes associated with gestation.

E. Labor

6. Explain the events associated with the three stages of labor.

F. Prenatal Diagnostic Tests

7. Describe several prenatal diagnostic tests such as amniocentesis, chorionic villus sampling (CVS), and fetal ultrasonography.

G. Exercise and Pregnancy

H. Key Medical Terms Associated with Developmental Anatomy

SCIENTIFIC TERMINOLOGY

Find an anatomical sample word for each prefix and suffix:

Prefix/Suffix	Meaning	Sample Word
allas-	sausage	
amnio-	amnion	
-bryein	grow	
-derm	skin	
ektos-	outside	
fertilis-	fruitful	
gastrula-	little belly	
-kentesis	puncture	
puer-	child	
syn-	joined	
teratos-	monster	
-tokos	birth	
troph-	nourish	
zygosis-	a joining	

A. Development During Pregnancy   (pages 811–815)

A1.   Check your understanding of fertilization by completing these questions.

a.   Fertilization usually occurs within _____ to _____ hours after ovulation.

b.   Due to the viability of the sperm and secondary oocyte, there typically is a (*2? 3? 4?*)-day window during which pregnancy can occur.

c.   The function of the enzyme _____, which is produced by the acrosome, is to (*stimulate? inhibit?*) sperm motility.

d.   Sperm cells must remain in the female reproductive tract for several hours, undergoing functional changes necessary for fertilization to occur. What change takes place during this period called capacitation?

e.   The acrosomal enzymes aid the sperm in penetrating the _____

_____, a ring of cells around the oocyte, and a clear glycoprotein layer

called the _____ _____.

f.   An ionic change within the membrane of the oocyte prevents _____, fertilization by more than one sperm cell.

g.   A _____ is a fertilized ovum, consisting of a segmentation nucleus, cytoplasm, and zona pellucida.

A2. Test your understanding of twins by completing the questions below. (Some questions have 2 answers.)

CT	conjoined twins
DT	dizygotic twins
MT	monozygotic twins

a. _____ Derive from a single fertilized ovum.

b. _____ Monozygotic twins that share skin, limbs, trunks, and viscera.

c. _____ Also known as fraternal twins.

d. _____ Are produced by the independent release of two secondary ova and fertilization by two different sperm.

e. _____ Referred to as identical twins.

A3. Complete the following exercise pertaining to morula formation, blastocyst development, and implantation.

a. The early mitotic divisions are called _____. The first division begins approximately *(12? 24? 36?)* hours after fertilization.

b. Three full days after conception there are *(8? 16? 32?)* cells. Each cell is referred to as a

_____.

c. The solid mass of cells, surrounded by the zona pellucida, produced a few days after

fertilization is called the _____.

d. By the time the morula enters the uterine cavity it has transformed into a hollow ball of

cells, a _____. This structure is divided into an outer covering of cells

called the _____ and an inner cell mass from which the

_____ will develop.

e. As this hollow ball of cells enters the uterine cavity it will remain free for a short period before attaching to the wall. During this time the *(corona radiata? zona pellucida?)* disintegrates.

f. The early secretions of the uterine wall that nourish the blastocyst are sometimes called

uterine _____.

g. It takes about _____ days from fertilization to implantation.

h. The portion of the trophoblast that contains no cell boundaries is the *(cytotrophoblast? syncytiotrophoblast?)*.

i. Implantation most commonly occurs on the *(anterior? posterior?)* wall of the fundus or body of the uterus.

A4. The development of an embryo or fetus outside the uterine cavity is referred to as an

_____ _____.

B. Embryonic Development (pages 815–821)

B1. The embryonic period is the first _____ _____ of development.

B2. Following implantation, the inner cell mass differentiates into the three primary germ

layers: _____, _____, and _____.

The movement of the cells to create these layers is known as _____.

B3. Match the embryonic event with the approximate day after fertilization that it occurs.

8th day	12th day	14th day

a. _____ Cells of the embryonic disc differentiate into three distinct layers: upper ectoderm, middle mesoderm, and inner endoderm.

b. _____ The cells of the inner cell mass proliferate and form the amnion and a space, the amniotic cavity, over the inner cell mass.

c. _____ Cells of the endodermal layer have been dividing rapidly, forming the yolk sac, another fetal membrane.

B4. The embryonic membranes lie (inside? outside?) the embryo. They function to

_____ and _____ the embryo and, later, the fetus.

B5. Match the embryonic membrane with its description.

A	amnion	C	chorion
AL	allantois	Y	yolk sac

a. _____ Derived from the trophoblast of the blastocyst and the mesoderm that lines the trophoblast. It will become the principal part of the placenta.

b. _____ Vascular outpouching of the hindgut; later its blood vessels serve as a connection between the mother and fetus (umbilical cord).

c. _____ Thin protective membrane that overlies the embryonic disc and eventually surrounds the embryo.

d. _____ Membrane that is the origin of blood vessels that transport nutrients to the embryo.

B6. What is the function of the amniotic fluid?

B7. The placenta is formed by a portion of the endometrium called the decidua

_____ and the chorion of the embryo.

B8. List the primary functions of the placenta.

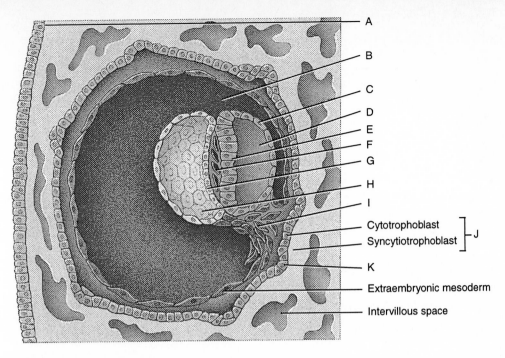

	A
	B
	C
	D
	E
	F
	G
	H
	I
	Cytotrophoblast ⎤
	Syncytiotrophoblast ⎦ J
	K
	Extraembryonic mesoderm
	Intervillous space

Figure LG 27.1   Formation of the primary germ layers and associated structures, internal view (about 14 days after fertilization).

B9.   Contrast the decidua basalis, decidua capsularis, and decidua parietalis.

B10.   Refer to Figure LG 27.1 and identify the following structures.

_____ amnion   _____ endoderm

_____ amniotic cavity   _____ endometrium of uterus

_____ body stalk (future umbilical cord) _____ extraembryonic coelom

_____ chorion   _____ mesoderm

_____ chorionic villus   _____ yolk sac

_____ ectoderm

B11.   Chorionic villi, which contain the fetal blood vessels, grow into the decidua basalis until

they are bathed in maternal blood sinuses called _____

_____.

B12.   Waste products from the fetus travel through umbilical *(veins? arteries?)* to the placenta and diffuse into the maternal blood.

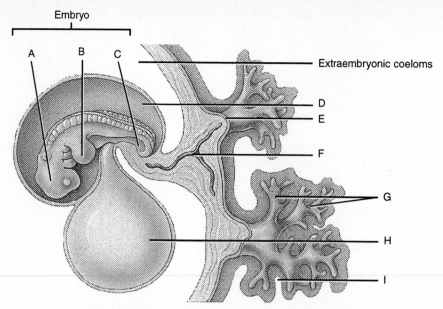

Embryo

A   B   C

Extraembryonic coeloms

D

E

F

G

H

I

Figure LG 27.2   Formation of the primary germ layers and associated structures, external view (about 25 days after fertilization).

B13.   Refer to Figure LG 27.2 and identify the following structures.

_____ amniotic cavity   _____ heart

_____ body stalk   _____ intervillous space

_____ chorion   _____ tail

_____ chorionic villi   _____ yolk sac

_____ head

C. Fetal Growth   (pages 821–823)

C1.   List some representative changes in the embryo and/or fetus that occur by the end of the first and third months. (Note: See Table 27.2.)

C2.   Discuss an example of fetal surgery.

D. Structural and Functional Changes During Pregnancy   (page 823)

D1.   The total human gestation period is about *(36? 38? 40?)* weeks.

D2.   List several anatomical changes associated with pregnancy.

D3. List several physiological changes associated with pregnancy.

D4. During pregnancy, the maternal heart rate increases by _____–_____% and the

blood volume by _____–_____%.

D5. During pregnancy the total body oxygen consumption can *(increase? decrease?)* by 10–20%.

D6. Define these skin changes associated with pregnancy.

a. Chloasma

b. Linea nigra

c. Striae

E. Labor   (pages 823–826)

E1. The onset of labor is accompanied by a(n) *(increase? decrease?)* in estrogen levels as

compared to progesterone levels. In addition, oxytocin from the _____
pituitary gland also stimulates uterine contractions, relaxes the pubic symphysis, and
helps to *(constrict? dilate?)* the cervix.

E2. A reliable indicator of true labor is the discharge of a blood-containing mucus plug called

the "_____."

E3. List the three stages of labor.

a.

b.

c.

E4. Complete dilation of the cervix to _____ cm occurs during the first stage of labor. The

term "afterbirth" refers to the delivery of the _____.

E5. Define puerperium.

E6. Reduction in the size of the uterus is called _____. The uterine
discharge, initially of blood and later of serous fluid, lasting up to 2–4 weeks after

delivery is known as _____.

F. Prenatal Diagnostic Tests   (pages 826–827)

F1. The removal of 10 ml of amniotic fluid at the 14–16th week of gestation is known as

_____. The test can detect about *(200? 300?)* chromosomal disorders and over *(30? 40? 50?)* biochemical defects.

F2. This diagnostic test utilizes an instrument that emits high-frequency sound waves that echo off the developing fetus back to a transducer that converts them into an image:

_____ _____.

F3. Name two advantages of chorionic villus sampling over amniocentesis.

G. Exercise and Pregnancy   (page 827)

G1. Discuss some effects of pregnancy on the ability to exercise.

ANSWERS TO SELECT QUESTIONS

A1. (a) 12, 24; (b) 3; (c) acrosin, stimulate; (e) corona radiata, zona pellucida; (f) polyspermy; (g) zygote.

A2. (a) CT, MT; (b) CT; (c) DT; (d) DT; (e) MT.

A3. (a) Cleavage, 24; (b) 16, blastomere; (c) morula; (d) blastocyst, trophoblast, embryo blast; (e) zona pellucida; (f) milk; (g) 6; (h) syncytiotrophoblast; (i) posterior.

A4. Ectopic pregnancy.

B1. Two months.

B2. Ectoderm; endoderm; mesoderm; gastrulation.

B3. (a) 14th; (b) 8th; (c) 12th.

B4. Outside, protect, nourish.

B5. (a) C; (b) AL; (c) A; (d) Y.

B7. Basalis.

B10. A—Endometrium of uterus, B—extraembryonic coelom, C—amnion, D—amniotic cavity, E—ectoderm, F—mesoderm, G—endoderm, H—yolk sac, I—body stalk (future umbilical cord), J—chorion, K—chorionic villus.

B11. Intervillous spaces.

B12. Arteries.

B13. A—Head, B—heart, C—tail, D—amniotic cavity, E—chorion, F—body stalk, G—chorionic villi, H —yolk sac, I—intervillous space.

C2. A team of surgeons removed a 23-week-old fetus from its mother's uterus, operated to correct a blocked urinary tract, and then returned the fetus to the uterus.

D1. 38.

D4. 10–15, 30–50.

D5. Increase.

E1. Increase, posterior, dilate.

E2. Show.

E3. Stage of dilation, stage of expulsion, placental stage.

E4. 10, placenta.

E6. Involution, lochia.

F1. Amniocentesis, 300, 50.

F2. Fetal ultrasonography.

F3. It can be performed as early as eight weeks of gestation; procedure does not require penetration of the abdomen.

SELF QUIZ

Choose the one best answer to the following questions.

1. Acrosin is secreted by the _____.

   A. ovaries
   B. anterior pituitary gland
   C. secondary occyte
   D. sperm
   E. placenta

2. Functional changes that sperm undergo, allowing them to fertilize a secondary oocyte, are referred to as

   A. syngamy
   B. polyspermy
   C. capacitation
   D. penetration
   E. implantation

3. Episodes of nausea and possible vomiting most likely to occur in the morning are called

   A. hyperemesis gravidarum
   B. emesis gravi-darum
   C. postpartum emesis
   D. emesis partum postulum

4. Implantation usually occurs about _____ after fertilization.

   A. 24 hours
   B. 72 hours
   C. 6 days
   D. 14 days
   E. none of the above are correct

5. Which of the following is NOT a symptom or sign of pregnancy?

   A. increased heart rate
   B. increased blood volume
   C. decreased oxygen consumption
   D. increased total body water
   E. difficult breathing

6. By the end of which month of embryonic and fetal development would urine start to form?

   A. 1st month
   B. 2nd month
   C. 3rd month
   D. 4th month
   E. 5th month

7. Which embryonic membrane provides an early site of blood formation?

   A. chorion
   B. allantois
   C. amnion
   D. yolk sac
   E. A and B are both correct

8. Which primary germ layer is responsible for producing all nervous tissue?

   A. ectoderm
   B. endoderm
   C. mesoderm
   D. B and C are both correct
   E. none of the above are correct

9. The portion of the endometrium that overlies the embryo, between the embryo and the uterine cavity, is the _____.

   A. decidua parietalis
   B. decidua fetalis
   C. decidua basalis
   D. decidua capsularis
   E. none of the above are correct

10. Which structure(s) is/are NOT derived from the mesoderm?

    A. skeletal muscle
    B. epithelium of the adrenal cortex
    C. cartilage, bone, and other connective tissue
    D. epithelium of the urinary bladder, gallbladder, and liver
    E. blood, bone marrow, and lymphoid tissue

Answer (T) True or (F) False to the following questions.

11. _____ Implantation of the placenta in the lower portion of the uterus, near or over the internal os, is referred to as placenta previa.

12. _____ Chorionic villus sampling requires penetration of the abdomen.

13. _____ The four- to six-week period following delivery of the baby and the placenta, during which time the reproductive organs and maternal physiology return to the prepregnancy state, is known as the puerperium.

14. _____ By the end of the third month, fine hair called lanugo covers the fetus's body.

15. _____ Cleavage produces an increase in both the number and size of the cells.

16. _____ Expulsion of the "afterbirth" refers to the delivery of the placenta.

17. _____ Dizygotic twins will always be the same sex.

Arrange the answers in the correct order.

18. Development stages _____ _____ _____ _____ _____

    A. embryo
    B. fetus
    C. zygote
    D. morula
    E. blastocyst

19. From outside to inside of an oocyte. _____ _____ _____ _____

    A. cytoplasm
    B. zona pellucida
    C. corona radiata
    D. plasma membrane of oocyte

Fill in the blanks.

20. Research indicates that in underweight or very lean females, the secretion of _____

    _____ _____ is abnormal in quantity and timing.

21. The early divisions of the zygote are called _____.

22. The portion of a blastocyst that is responsible for the development of the membranes composing the placenta is

    the _____.

23. _____ is a syndrome characterized by sudden hypertension, large amounts of protein in the
    urine, and generalized edema. It may be an autoimmune or allergic reaction to the fetus.

24. The malpresentation of the fetal buttocks or lower extremities into the birth canal is known as a

    _____ presentation.

25. _____ _____ is a disease resulting from an infection originating in the
    birth canal that later affects the endometrium. It is sometimes called childbed fever.

ANSWERS TO THE SELF QUIZ

1. D
2. C
3. B
4. C
5. C
6. C
7. D
8. A
9. D

10. D
11. T
12. F
13. T
14. F
15. F
16. T
17. F
18. C, D, E, A, B

19. C, B, D, A
20. Gonadotropin-releasing hormone
21. Cleavage
22. Trophoblast
23. Preeclampsia
24. Breech
25. Puerperal fever